Ethics for Radiation Protection in Medicine

Series in Medical Physics and Biomedical Engineering

Series Editors

John G. Webster, E. Russell Ritenour, Slavik Tabakov, and Kwan Hoong Ng

Recent books in the series:

Handbook of X-ray Imaging: Physics and Technology
Paolo Russo (Ed)

Quantitative MRI of the Brain: Principles of Physical Measurement, Second Edition
Mara Cercignani, Nicholas G. Dowell, Paul S. Tofts (Eds)

Graphics Processing Unit-Based High Performance Computing in Radiation Therapy
Xun Jia, Steve B. Jiang (Eds)

A Guide to Outcome Modeling In Radiotherapy and Oncology: Listening to the Data
Issam El Naqa (Ed)

Radiotherapy and Clinical Radiobiology of Head and Neck Cancer
Loredana G. Marcu, Iuliana Toma-Dasu, Alexandru Dasu, Claes Mercke

Advances in Particle Therapy: A Multidisciplinary Approach
Manjit Dosanjh, Jacques Bernier (Eds)

Problems and Solutions in Medical Physics: Diagnostic Imaging Physics
Kwan Hoong Ng, Jeannie Hsiu Ding Wong, Geoffrey D. Clarke

Advances and Emerging Technologies in Radiation Oncology Physics
Siyong Kim, John Wong

Clinical Radiotherapy Physics with MATLAB: A Problem-Solving Approach
Pavel Dvorak

For more information about this series, please visit: https://www.crcpress.com/Series-in-Medical-Physics-and-Biomedical-Engineering/book-series/CHMEPHBIOENG

Ethics for Radiation Protection in Medicine

Jim Malone
Friedo Zölzer
Gaston Meskens
Christina Skourou

CRC Press is an imprint of the
Taylor & Francis Group, an informa business

CRC Press
Taylor & Francis Group
6000 Broken Sound Parkway NW, Suite 300
Boca Raton, FL 33487-2742

Printed on acid-free paper

International Standard Book Number-13: 978-1-1385-5388-0 (Hardback)

Visit the Taylor & Francis Web site at
http://www.taylorandfrancis.com

and the CRC Press Web site at
http://www.crcpress.com

Contents

About the Series

THE SERIES IN MEDICAL PHYSICS AND BIOMEDICAL ENGINEERING describes the applications of physical sciences, engineering, and mathematics in medicine and clinical research.

The series seeks (but is not restricted to) publications in the following topics:

- Artificial organs
- Assistive technology
- Bioinformatics
- Bioinstrumentation
- Biomaterials
- Biomechanics
- Biomedical engineering
- Clinical engineering
- Imaging
- Implants
- Medical computing and mathematics
- Medical/surgical devices
- Patient monitoring
- Physiological measurement
- Prosthetics
- Radiation protection, health physics, and dosimetry
- Regulatory issues
- Rehabilitation engineering
- Sports medicine
- Systems physiology
- Telemedicine
- Tissue engineering
- Treatment

The Series in Medical Physics and Biomedical Engineering is an international series that meets the need for up-to-date texts in this rapidly developing field. Books in the series range in level from introductory graduate textbooks and practical handbooks to more advanced expositions of current research.

The Series in Medical Physics and Biomedical Engineering is the official book series of the International Organization for Medical Physics.

THE INTERNATIONAL ORGANIZATION FOR MEDICAL PHYSICS

The International Organization for Medical Physics (IOMP) represents over 18,000 medical physicists worldwide and has a membership of 80 national and 6 regional organizations, together with a number of corporate members. Individual medical physicists of all national member organisations are also automatically members.

The mission of IOMP is to advance medical physics practice worldwide by disseminating scientific and technical information, fostering the educational and professional development of medical physics and promoting the highest quality medical physics services for patients.

A World Congress on Medical Physics and Biomedical Engineering is held every three years in cooperation with International Federation for Medical and Biological Engineering (IFMBE) and International Union for Physics and Engineering Sciences in Medicine (IUPESM). A regionally based international conference, the International Congress of Medical Physics (ICMP) is held between world congresses. IOMP also sponsors international conferences, workshops and courses.

The IOMP has several programmes to assist medical physicists in developing countries. The joint IOMP Library Programme supports 75 active libraries in 43 developing countries, and the Used Equipment Programme coordinates equipment donations. The Travel Assistance Programme provides a limited number of grants to enable physicists to attend the world congresses.

IOMP co-sponsors the *Journal of Applied Clinical Medical Physics.* The IOMP publishes, twice a year, an electronic bulletin, *Medical Physics World.* IOMP also publishes e-Zine, an electronic news letter about six times a year. IOMP has an agreement with Taylor & Francis for the publication of the Medical Physics and Biomedical Engineering series of textbooks. IOMP members receive a discount.

IOMP collaborates with international organizations, such as the World Health Organisations (WHO), the International Atomic Energy Agency (IAEA) and other international professional bodies such as the International Radiation Protection Association (IRPA) and the International Commission on Radiological Protection (ICRP), to promote the development of medical physics and the safe use of radiation and medical devices.

Guidance on education, training and professional development of medical physicists is issued by IOMP, which is collaborating with other professional organizations in development of a professional certification system for medical physicists that can be implemented on a global basis.

The IOMP website (www.iomp.org) contains information on all the activities of the IOMP, policy statements 1 and 2 and the 'IOMP: Review and Way Forward' which outlines all the activities of IOMP and plans for the future.

Preface

DURING THE FINAL STAGES of the preparation of the manuscript for this book, two new vital related documents were published. They were, in order of appearance, the revised Declaration of Geneva from the World Medical Association (WMA) which was published in October 2017, and the International Commission on Radiological Protection's (ICRP) Report 138: *Ethical foundations of the system of radiological protection*, which appeared in early 2018. Both are documents of great importance for ethics and radiation protection in medicine. The former, a modern-day version of the Hippocratic Oath, emphasises, among other values, the importance of the autonomy and dignity of the individual, which has not to date been a central consideration in the radiation protection of patients. The latter identifies the ethical values that underpin the system of radiation protection in general, for example in dealing with the Fukushima disaster. It defers to a further report in the early 2020s the value framework for the ethics issues specific to radiation protection in medicine.

Neither of these documents offers a definitive solution to the ethical issues for radiological protection in medicine, yet both are helpful and reassuring in terms of the approach adopted in this book. This is a practical, pragmatic approach developed in Chapters 2 and 3, and reaches the conclusion that the values required for radiation protection of the patient in medicine, the focus of this book, include:

- Dignity and autonomy of the individual;
- Non-maleficence/beneficence (i.e. do no harm and do good);
- Justice;
- Prudence/precaution; and
- Honesty/transparency (particularly in presenting information to patients).

The basis for this pragmatic value set was published in its present form in 2016 and overlaps with both the Geneva Declaration and ICRP 138 (Malone and Zolzer 2016; ICRP 2018; Parsa-Parsai 2017).

Almost 20 years earlier, significant ethical questions were emerging in medical radiation protection that seemed outside the provenance of the three ICRP principles as then articulated. These first received public attention in several projects undertaken for the European Commission (EC) in support of implementation of their then recently issued MED (Medical Exposures Directive) (EC 1997). These included actions in the DIMOND and SENTINEL projects, which were concerned with practices in diagnostic imaging. In 2006–2007, some of the ethics issues involved were brought together in two EC conferences on radiation protection in digital imaging: a full session at the SENTINEL conference in Delft in 2007, and a two-day workshop in Dublin in 2006 (Faulkner et al. 2008; Malone et al. 2009). In addition, other publications in the EC Radiation Protection series highlighted issues in which ethical sensibility was an important factor, for example, one on clinical audit of radiology and one on clinical guidelines for when imaging is appropriate (EC 2000, 2008, 2009, 2018).

These initiatives, driven by a concern with justification in medicine, became an important focus for the International Atomic Energy Agency (IAEA) between 2007 and 2012. Related ethics exploration identified systemic deficits in practice. Perhaps surprisingly, the deficits noted included problems with *respect for individual autonomy*; *do no harm*; *the need for openness, transparency and accountability*; and *equitable distribution of resources and risk* among others. Later work also emphasised the *precautionary principle* and *prudence* as important in managing the inevitable uncertainties involved (IAEA 2011).

The opportunity for these strands to come together first presented itself at chance meetings of two of the authors in 2007, at an ethics session during an International Radiation Protection Association (IRPA) conference in Brasov, and later at meetings during 2012–2013 in Milan and Stockholm. These encounters led to a publication proposing the pragmatic set of values for medical radiation protection (Malone and Zolzer 2016). The circumstances leading to the involvement of the other authors were equally serendipitous.

The evolution of the explicit ethics framework for radiation protection as detailed in ICRP 138 reached conclusions similar to the pragmatic set, in terms of the values to which it attributed importance: surprisingly, since ICRP and the pragmatic value study had different objectives and followed different trajectories. The latter was concerned with identifying the values essential to the day-to-day practice of good radiation protection in medicine. The ICRP report had, on the other hand, the objective of excavating the ethical basis of the system of radiological protection as a whole and followed a broadly based consultation process. The system was under exceptional pressure at the time. Unsurprisingly, perhaps, at a meeting in Fukushima City in 2012, ICRP began to examine the ethics issue, led by Jacques Lochard. A task group was formed in October 2013, initially chaired by Professor Deborah Oughton, and later by Kun-Woo Cho. Its report eventually became ICRP 138 discussed above, a great advance that is frequently referred to in this book.

Notwithstanding, while the systems of medical ethics and the ICRP system for radiation protection overlap significantly, there is no simple way of mapping one onto the other. This will be a task for another day. However, as the initiatives leading to the pragmatic set matured, it became obvious that an important unstated aspect of the medical system of radiation protection is that medical actions, including the radiological, must be performed in a way that is consistent with an established system of medical ethics. Beauchamp's and Childress' approach provides a ready starting point to fulfil such a requirement. Surprisingly, some of us had already been exploring their values individually without thinking through their connectedness with a system of medical ethics. In retrospect was an obvious insight to miss.

When it comes to making sense of technological risk, such as the radiation risk in medical imaging or radiotherapy, or in the use of fossil fuels, or in numerous medical/pharma dilemmas, the position of science is complicated. On the one hand, the pressure on science to deliver evidence in the service of politics, medicine, society, or the market is higher than ever before. On the other, we realise that, in many circumstances, science does not, and possibly never will, produce the full rationale for risk acceptance (or rejection). This dilemma applies to numerous areas including nanotechnology, mobile phones, pharmaceuticals, genetically modified organisms and many others. Today, scientific hypotheses acquire a medical, social and political function. They are released from the laboratory, without full support from empirical evidence, but with a political, social, industrial or medical task to accomplish. From this we learn that the uncertainties which plague some aspects of radiation protection of patients are not unique. But, our dogged insistence that the only solution worth looking at is in the sciences is, if not unique, seriously problematic (see also the afterword).

Where there is persistent uncertainty in the sciences, including biomedical sciences, we must look to ethics for guidance. Lamenting the uncertainty is not a productive path to take: like the poor, it will always be with us, in some aspects of science and medicine. Clearly, this is at odds with the traditional understanding of the science, medicine, and ethics spectrum. In practice, the main product of science, including the biomedical sciences, is no longer empirical evidence to prove or reject a specific hypothesis, but the hypothesis itself. Thus, we need a broader, more deliberative conception of science, and the ethical framework within which it works. Science and (or) medicine cannot and should not do that alone. It should be done interactively with the relevant community, including doctors, patients, scientists, other health-care professionals, politicians, managers, citizens and activists.

Jim Malone
Friedo Zölzer
Gaston Meskens
Christina Skourou
RICOMET Conference
Antwerpen, Belgium, 2018

Authors

Jim Malone is Robert Boyle Professor (Emeritus) of medical physics at Trinity College Dublin. He worked halftime with the International Atomic Energy Agency (IAEA) (Vienna) for several years and continues as a consultant with both the agency and WHO. He was dean of the School of Medicine and Faculty of Health Sciences at Trinity College Dublin and chairman of the Geneva-based International Electrotechnical Commission's (IEC) committee for global standards for medical imaging equipment. He has over 300 publications and numerous conference papers/keynote addresses. He has wide interests in the humanities and has contributed to thinking on the ethics basis for radiation protection in medicine. He has been involved in organising numerous summer schools involving science, medicine and the humanities.

Friedo Zölzer is professor and head of the Institute of Radiology, Toxicology and Civil Protection at the Faculty of Health and Social Sciences of the University of South Bohemia in České Budějovice, Czech Republic. He is a radiobiologist with a strong interest in ethics, regularly lecturing and publishing in the area. He has significantly contributed to the development of a new approach to ethics of radiation protection at the international level, among others as a member of the International Commission on Radiological Protection's (ICRP) task group on 'Ethics of Radiological Protection' and has been the organiser of a widely recognised biennial series of international symposia on the ethics of environmental health, in which medical applications for radiation played an important part.

Gaston Meskens is a theoretical physicist with over 15 years' experience of research related to sustainable development, energy, climate change and radioactive waste management. He works part-time with the Centre for Ethics and Value Inquiry of the Faculty of Arts and Philosophy of the

University of Ghent and with the Science and Technology Studies group of the Belgian Nuclear Research Centre SCK•CEN . His research focusses on a human rights perspective on intellectual capacity-building in the interest of global sustainable development governance. At SCK•CEN, he co-founded the Programme of Integration of Social Aspects into Nuclear Research in 1999. The programme takes nuclear technology as a case, in order to critically study the complexity of risk-inherent technology assessment from the perspective of social justice and sustainable development. Recently, he has been a key contributor to the ethics framework for radiation protection for the International Commission on Radiological Protection (ICRP). At SCK•CEN, he works as researcher, writer, lecturer and mediator of dialogue on ethics in relation to science and technology.

Christina Skourou, is a clinical radiotherapy physicist at the St. Luke's Radiation Oncology Network in Dublin, Ireland, where she is involved in numerous radiotherapy clinical trials and is a member of the network's Research Ethics Committee. She has served as the chair of the American Association of Physicists in Medicine (AAPM) Task Group 109, revising the Code of Ethics and, as of January, 2019, is the chair of the AAPM Ethics Committee. Christina holds a PhD in biomedical engineering (Dartmouth College) and an MA in medical ethics and law (Queen's University Belfast).

CHAPTER 1

Introduction

1.1 INTRODUCTION

Radiation is a significant positive contributor to modern healthcare and is used for both diagnostic and therapeutic purposes. The diagnostic applications are for the greater part in medical imaging, within which the most frequently used modality is radiology. For example, multislice computed tomography is a highly successful and widely available technique. The therapeutic interventions are mainly, though not exclusively, in radiation oncology. Both have greatly enhanced the effectiveness of medical practice and have overseen technology transfer on an exceptional scale into the healthcare system, in a relatively short time (ICRP 2007b).

The World Health Organisation (WHO) definition of health as a state of complete physical, mental and social well being is broad, including ethical, social, public health and resource considerations (WHO 2006). Accepting this definition requires that we look at the societal background to the radiological protection system as part of establishing an ethical framework suited to its application in medicine. The radiation protection system is based on a mixture of scientific evidence, practical experience and value judgements and has an ethical dimension that until recently has not been explicitly stated (ICRP 2018). In medicine, the principles of the International Commission on Radiological Protection (ICRP) (i.e. justification, optimisation and dose limitation) must be implemented in light of the values generally informing medical practice and ethical behaviour in society.

However, the benefits of radiation come at a price, and its range of application now accounts for over 98% of man-made human radiation exposure, which is accompanied by an increased population radiation burden and associated probable risks. Every day more than 10 million diagnostic procedures are performed globally, amounting to some 3–4 billion annually (ICRP 2007b; UNSCEAR 2012). Similarly, 7 million cancer patients may benefit from radiotherapy globally each year (Jaffray and Gospodarowicz 2015). The increase in dose and risk can be acceptable when a real benefit flows from it, such as improved diagnosis, a better treatment outcome, or better management of the patient. However, this is not always the case and patients can receive significant exposures, without receiving any information, and sometimes without commensurate benefit. For many years, diagnostic radiation dosage was regarded by many as a non-issue, but this is no longer acceptable (NCRP 2009).

1.2 SOCIETAL ISSUES AND THE HEALTH SECTOR

Medical practice and medical imaging necessarily take place within the context of the general developments in society and the expectations of its citizens. This, and its consequences for radiation protection in the medical sphere, have been reviewed at some length elsewhere. A few of the more important points are summarised here. First, it is important to be aware of the scale of resource allocation to healthcare, consuming as it stands 10%–20% of national budgets in many countries (Papanicolas et al. 2018). This alone would set it aside as an area of special concern from both practical and ethics viewpoints. Here, it is important to note that the historical paternalism of the medical professions no longer provides an acceptable approach to service delivery and interpersonal behaviour within the services. There are many other shifts in basic values and social concerns.

Here is a short list of societal areas in which there has been profound change since the principles on which the ICRP operates (ICRP 1966, 1977, 1991) were formulated and introduced: marriage, divorce, single parents, gay marriage, disability rights, gender rights, distrust of authority/professions, the right to life, euthanasia/assisted suicide in some jurisdictions, the dominant presence of social media, and widespread acceptance of the right to privacy/autonomy of the individual. In many cases, these changes are reflected in the law, social policy, and practices of society. This is particularly so in medicine where there have also been substantial shifts in practice, often driven by social or legal developments and often initially resisted by the health professions. In particular, there have been significant developments around the areas of patient status and consent (Malone 2008, 2009; Malone et al. 2009). It is thus evident that the principles of radiation protection must be applied in a healthcare system shaped by social forces that had little impact a few decades ago.

While the spend on healthcare is greater than before, the impact of health economics and special interest groups on decision making has become very important and can, on occasions, override real medical priorities and individual clinical decisions. Interest groups may divert resources to benefit their group; health professionals may be under pressure to optimise revenue; bureaucracies, including regulatory agencies, can be self-serving to the detriment of common good; and politicians need to deliver for the public at large (e.g. a local mammography screening programme). These problems and many more have given rise to a more formal approach to Health Technology Assessment (HTA). Arising from such studies there is now, in the wider literature, a strong evidence base for the view that in some countries there is significant over-utilisation of imaging (IAEA 2011; Papanicolas et al. 2018) and new radiotherapy technologies (Curry et al. 2014; Hager et al. 2015).

In addition, there is the ever-present issue among the public and other professions, of the risk(s) of radiation, real or imagined. The starting point for such discussion must be that radiation is a known carcinogen. How this impacts specific situations in diagnostic imaging or radiotherapy is the subject of the following chapters. For the moment, some brief comments are in order. The American College of Radiology (ACR) 'White paper on Radiation Dose in Medicine' suggests that current imaging rates may result in an increased incidence of radiation-related cancer in the near future. Some estimate the increase could be up to 1.5% or 2% (BEIR VII 2006; Amis et al. 2007). Of course, if the scan is necessary in serious or life-threatening situations, it must be done. (Malone et al. 2012; Semelka et al. 2012).

Estimates of deaths and future cancers in many publications hide notable uncertainty about their origins, significance and how they might be presented to patients and other health professionals. For example, the importance of a risk of a few per cent of deaths occurring 10 years into the future will be seen very differently by a young mother and by an octogenarian man with multiple pathologies. Likewise, there are great differences between the way risks are calculated and the way the benefits are estimated; frequently, it is a matter of apples and oranges, rendering it almost impossible to do real risk benefit estimates. Similar considerations may apply to comparisons with risks and benefits from other treatments/procedures and/or medications. We will return to these topics time and time again throughout the book.

Another important related area in which the prevailing environment has radically altered since the introduction of the present system of radiation protection in the 1960s, is in the openness, accountability, transparency and honesty that is now expected of medical professionals and the institutions in which they serve. This is obviously different in different parts of the world (Malone and Zölzer 2016; ICRP 2018). However, the direction in which external pressures are applied is invariably towards more openness, accountability and transparency. Likewise, in terms of the view of the person, there is now a high level of consensus in most political, social and legal systems respecting the dignity of individuals, their autonomy and their right to respect. All the above, and other obligations that will be discussed in later chapters, impose new burdens on professions not accustomed to this type of expectation and/or oversight (Malone and Zölzer 2016; Parsa-Parsi 2017).

Where these new expectations are not met and when, consequently things go seriously wrong in medicine in general, enquiry may initially follow a peer review-like process. In the past, where this failed, enquiry by a professional body often yielded acceptable results. However, it is now common for the findings of such a group to be regarded as unsatisfactory and self-serving. When this is the case, formal (often judicial tribunals of enquiry) follow to determine the pertinent matters of fact, which are sometimes pursued through the courts of law to confirm the facts and assign guilt/punishment. This has now become a common and accepted feature of the lives of health professionals (Malone et al. 2012).

1.3 CULTURE AND PROFESSIONS

The framework or 'culture' within which professionals operate may be considered from a point of view often taken by anthropologists, ethnographers or social scientists. In the nineteenth and early twentieth centuries, it was common for anthropologists to visit 'newly discovered' countries and/or tribes and report on the ways of life and the different cultures they encountered. This approach has been extended to subgroups of western society by social scientists, ethnographers and anthropologists.

We are familiar with way-of-life studies and culture of disadvantaged subgroups. However, similar methodology can be applied to any identifiable group to expose the culture within which it operates. The group might be, for example: clerics, doctors, software engineers, or other professionals, including radiation protection specialists (Malone et al. 2012).

In these studies, the term culture is much broader than implied when it is used to denote some aspects of the arts. Wilson (2008), in a study of the decline of a highly identifiable group (clergymen), describes culture as follows:

> [It] involves very concrete patterns of behaviour and ways of thinking that give shape to a particular body of people–whether we can put names on those features or not... It has its shape because of a deep and commonly held set of standards and expectations which come to expression in the behaviours of the collection of players... Living out a culture, with its innumerable assumptions and expectations, inevitably evokes in us a challenge when we come face to face with persons operating in a different one: we find it difficult to understand their behaviour because we don't know where it is coming from.
>
> The expected attitudes and behaviours of [those involved in] a particular culture can be so powerful that it becomes all but impossible for its members to even conceive of other ways of being.
>
> Finally, cultures cling to existence tenaciously, for at least two reasons... The first lies precisely in the fact that much of their causation is unacknowledged. The second... lies in its capacity to generate meaning... For the individual who risks acting out a different paradigm, the cost in terms of rejection by the players who want to continue with the reassuring story may be high.

These characteristics can be applied to many groups, including physicians, radiologists, radiation protection professionals, regulators and the general public. Each group has, to some extent, the characteristics described by Wilson and many other thinkers in the area. The individual may be a member of one or more of these groups and while functioning as a member of that group will adopt the norms and approaches of the group, i.e. will live according to the culture of the group. Obviously, there are great advantages in a profession in having a healthy, responsive culture. This has been recognised in radiation protection, and the IRPA have issued a guide to developing the culture of radiation protection, including its positioning in society and among all the relevant stakeholders (IRPA 2014). In terms of effectiveness, the law is helpful in achieving the objectives of radiation protection, but a good culture may well be better.

1.4 RADIATION IN MEDICINE

Modern medical practice is so multifaceted that it defies a comprehensive description. Perhaps one defining characteristic is its immense scientific and technological success coupled with an iconic repositioning in public consciousness. Instances of medical progress are too numerous to mention, but examples include: minimally invasive surgery, cardiac interventional procedures such as stenting and electrophysiological treatments, and pain medicine injection techniques often requiring significant radiation commitments. Such progress has been accompanied by a growth in the public expectation of hospitals and medical institutions, to a level that is probably unrealistic and places an undue burden on the system and those working in it. This also, inevitably, creates public disappointment and anger when expectations are not met (Malone 2008; Malone et al. 2009).

A simpler measure of the importance of medicine in the life of a country is, as already noted, the scale of investment in both financial and human terms. In socialised systems it can become one of the largest items of government expenditure, and a corresponding component of the working lives of a large fraction of the population. Investment on this scale can only occur when the community regards it as important. Indeed, it has been argued that it is an iconic activity in which the public invests much of its hope and its aspirations to care both for itself and for others, when such care is needed. Thus, modern healthcare is an important part of the culture in which we live our lives.

This view of medicine as an iconic activity in society is further attested to by the number of medical soap operas that appear on television (see Table 1.1). When human societies do not fully understand what is happening in an area, they often create and tell stories that carry some (or all) of the meaning that cannot be articulated in a more conventional context by management and policy makers. The position of medical television soaps reflects this deep characteristic, our flawed understanding of the healthcare system, and our expectations of it. The last entry in the table is the cult soap 'Green Wing' in which a radiologist is one of the lead

TABLE 1.1 Ten Popular Medical Soaps (from 2017 listings)

• Grey's Anatomy	• General Hospital
• Code Black	• Holby City
• ER	• The Doctors
• Casualty	• House
• Doc Martin	• Green Wing

Based on: Malone, J.F., *Radiat. Prot. Dos.*, 135, 71–78, 2009.

characters. Further light is thrown on this by both the engagement with art in modern hospital buildings and by the explicit targeting of health issues in some contemporary art, such as that by Damien Hirst.

From another perspective, the model for provision of medical services continues to harbour strong paternalist elements, while the public-ethics context within which it operates has changed radically. It is not uncommon to encounter evidence of desensitisation of professions to the concerns of the public. They sometimes fail to recognise that growth in individual autonomy, consumerist culture, transparency and accountability are dominant influences in the way social (including medical) transactions are expected to take place. Failures in these areas have led to distrust of the authority of professions, and have ultimately, in some countries, led to the collapse of professional self-regulation (GMC 2008). Examples of these phenomena can be studied in the history of various medical scandals, such as: the fatal problems with blood products, the Harold Shipman Enquiry in the UK, the infant organ retention issues and many others (National Archives 2009). Table 1.2 lists some of the major medical scandals of the last few decades, which challenged the professions involved to revise their approach to ethics issues. It has become obvious following investigations and/or public enquiries into such events that a gap has opened up between what is acceptable to the public on one hand, and what appears reasonable to, or at least is accepted by the professionals involved, on the other.

TABLE 1.2 When Medical Practice Model Fails

• Shipman deaths in the UK
• Blood Products Contamination
• Retention of Infant Organs Postmortem
• Collapse of Self-regulation
• The Dawn Raid – A Salutary Incident

Based on: Malone, J.F., *Radiat. Prot. Dos.*, 135, 71–78, 2009.

The last entry in Table 1.2, The Dawn Raid, is an oblique reference to the early morning arrest, at their homes, of two senior, well-established medical professionals, one the medical director of a national blood transfusion service, and the other, chief biochemist in the same service. They were then charged with serious crimes that resulted in patients dying. What was unusual about this is that both are/were well regarded by their peers within the culture of their professions. Many were shocked by the charges, and felt that both had conducted themselves according to the standards the profession expected. However, public enquiry and formal judicial investigation did not reach similar conclusions and were critical of both. The charges involved were eventually not pursued. This is not a unique occurrence and similar shock has characterised professional reactions to the conclusions reached by public enquiries into the action of medical professionals; an example of this is the initial response of pathologists to the public reaction to judicial enquiry into postmortem retention by hospitals of infants' internal organs (DOHC 2006; Malone 2008, 2009; Malone et al. 2009). Over and above these essentially systemic problems, litigation and distrust of authority, which are growing features of medicine are, at least in part, due to paternalism and failure of physicians to respect the individual autonomy of patients and to communicate with and be accountable to them.

Radiology has struggled valiantly to establish its position within medicine, and has fostered its expertise to protect a constantly shifting position. It has made numerous commendable organisational and clinical contributions to a good radiological protection agenda; but there are also some weaknesses. One of the more important of these is the weakness of the justification process (Chapters 3 and 4) in much of diagnostic radiology.

A major difficulty in the post-modern era is getting engagement from scientists, doctors and managers in debates on ethical and related issues. In the English language this reflects what some see as trivialisation of debate in these areas, so that they appeared to be language games for much of the period between and following the two world wars. While the language games persist, accessible work is being produced that provides an excellent basis for ethical debate and education on radiation protection in medicine. Valuable and constructive contributions to the area are now also emerging from the social sciences, academic bioethics studies and the humanities (Malone and Zölzer 2016; ICRP 2018).

Public life must now respond to moves to individual autonomy. Medicine and medical services inevitably experience these influences. It is clear that the consumerist approach is already present in the language and patterns of provision even in some socialised medical systems. There is also evidence of changing models of access to medical facilities, particularly in medical imaging and radiotherapy. For example, medical tourism is encouraged by governments, industry and the professions and is now commonplace in many holiday destinations. In radiology, the growth throughout the world of freestanding private imaging clinics is widespread; the feeling among 'customers' of these clinics may be that, if they want an examination, they should be allowed to have it. Promotional websites, leaflets, brochures and other marketing material appear to encourage this.

1.5 SCOPE

The purpose of this book is to introduce the ethical values that are essential to the practice of radiation protection of the patient. It is designed to be relatively short, authoritative and accessible, rather than exhaustive.

A pragmatic set of five values, essential to the practice of radiation protection in radiology and radiotherapy is developed in Chapters 2 and 3. Three of these are fundamental to medical ethics and appear to have wide applicability and acceptability in all cultures. From the areas identified as needing ethical reflection, justification, communication of risk, prudence and honesty are probably the most obvious, but the others are also necessary. Chapter 6 extends the reflection to additional values that would support and improve the practice of radiation protection in medicine in the future.

The application of the value set is illustrated in Chapters 4 and 5, dealing with imaging and radiotherapy respectively. Ten clinical scenarios in each chapter illustrate how the values impact clinical situations in practice and provide concrete examples to help the reader visualise how their application might evolve. Chapter 7 broadens the perspective by looking beyond immediate concerns and examines the ethics–science relationship and addresses questions of heritage, sustainability, and impact of our approach into the future.

The intended audience includes radiologists, radiation oncologists and medical physicists; medical practitioners referring people to radiology or radiation oncology; technologists, radiographers, radiation therapists, nurses and other allied professionals; service managers in radiology and radiation oncology; participants in hospital ethics committees or institutional review boards; senior managements of health facilities; academics in undergraduate and postgraduate schools for any of the above; and academics/researchers in the sciences, humanities and social sciences.

CHAPTER 2

Ethics for Radiation Protection in Medicine

Framework and Multicultural Considerations

2.1 ETHICAL ASPECTS OF RADIOLOGICAL PROTECTION

The considerations of dose and risk discussed in the preceding chapter, be they medical or non-medical, suggest that radiological protection is not a matter of science alone, but clearly has ethical aspects. For example: Why do we find a dose limit of 20 mSv per year (i.e. a probable lifetime risk of fatality of 1:1,000) tolerable for those professionally exposed to radiation? Why do we have no dose limit for medical exposures? Why do some people prefer a computed tomography scan over a simpler radiography, even if the latter would be entirely sufficient for our purpose? Some of it comes down to economic considerations: reducing professional exposures is expensive, the computed tomography equipment has to be amortised, and so on. And the question then is how to weigh risks of human health against finances. But there are other factors: some people are not willing to give up what they are used to and do not care about assumed radiation risks; others are extremely apprehensive and do not want to accept a small risk even if they are assured the alternative is much riskier. There is also the question of how to act under uncertainty: How sure do we need to be about health risks before we adopt measures? And finally, there is the problem of how decisions are made: Should they be made by the specialists alone, or with the inclusion of all those affected?

The International Commission on Radiological Protection (ICRP) provides recommendations and guidance on all aspects of protection against ionising radiation. It might be expected to have questions like these addressed long ago, but until quite recently, it largely ignored them. Radiological protection was considered a matter of science, and perhaps of practical experience, but not of ethics. Gradually, it was realised that this understanding was too narrow, and that the system of radiological protection had implicitly referred to certain values from its very beginning. In 2013, it was finally decided to face this challenge. A task group on 'Ethics of Radiological Protection' was appointed and was asked to review the Commission's publications with respect to ethical aspects that might be contained therein.

At the same time, ICRP, together with the International Radiation Protection Association (IRPA), initiated a series of workshops around the world, two each in Europe, East Asia and North America, in which relevant topics were discussed with radiological protection professionals as well as ethics specialists. The task group also benefitted from discussions in a number of other symposia and congresses where special sessions were held on ethics of radiological protection. A draft report was compiled and after preliminary approval by the Commission, was offered for public consultation through the ICRP website. The final text was approved in 2017 and published 2018 (ICRP 2018).

The report first provides a historical analysis of the ICRP recommendations that endeavours to show that ethical values have been at the basis of the system of radiological protection throughout its evolution, although as mentioned previously, they were rarely made explicit. The document does not aim to present completely new ideas for the ethics of radiological protection, but rather to identify a minimum set of values relevant to the ICRP system in the past and present, and applicable to current and foreseeable problems with sensible results. The second part of the document then presents what are called the 'core values' underlying radiological protection as well as some 'procedural values' which are supposed to drive its practical applications.

The Core Values are:

- Beneficence/Non-maleficence: *Do good, and avoid doing harm*
- Prudence: *Recognise and follow the most reasonable course of action, even when full knowledge of its consequences is not available*
- Justice: *Distribute benefits and risks fairly*
- Dignity: *Treat individuals with unconditional respect, and recognise the capacity to deliberate, decide and act without constraint*

Under the heading 'procedural values', the document mentions

- Accountability: *Be answerable to all those who are possibly affected by your actions*
- Transparency: *Be open about decisions and activities that may affect others and communicate them in a clear, accurate, timely, honest and complete manner*
- Inclusiveness: *Involve all relevant parties (stakeholders) in decision-making processes*

2.2 COMPARISON WITH THE *PRINCIPLES OF BIOMEDICAL ETHICS*

As readers who are familiar with the *Principles of Biomedical Ethics* by Beauchamp and Childress (first edition 1979, latest edition 2013) will readily see the 'core values' have some similarity with that system, however, there are also differences.* Beauchamp and Childress present beneficence and non-maleficence as two independent values, whereas ICRP combines them into one, emphasizing the necessity of balancing good and harm. This reflects the first principle of radiological protection: justification. Justification is defined as *'Any decision that alters the radiation exposure situation should do more good than harm'* (ICRP 103 2007).

Prudence is not mentioned as an independent value by Beauchamp and Childress, but it obviously is of importance for radiological protection, where sometimes action has to be taken without exact knowledge of the risks

* Whereas Beauchamp and Childress speak of 'principles', the International Commission on Radiological Protection preferred the term 'core values' to distinguish ethical concepts from the three 'principles of radiological protection' (justification, optimisation, dose limitation).

involved, for instance at low doses. In such cases, the second principle of radiological protection applies: optimisation or '*The likelihood of exposure, the number of people exposed, and the magnitude of their individual doses shall be kept as low as reasonably achievable, taking into account economic and societal factors*'. (ICRP 103 2007) Here the underlying assumption is that even small doses of radiation carry some risk (see Chapter 1), which increases linearly with dose, and that this approach to protection is based on 'prudence'. Although ICRP also refers to 'precaution', this is not the only consideration guiding optimisation; 'economic and societal factor' are not to be neglected. So 'prudence', at least as used by ICRP, is a broader concept than 'precaution' because it looks at the benefits for the larger whole as well as at the risks of the individual. We will come back to this particular aspect in Chapter 6.

Beauchamp and Childress (1979) list respect for autonomy instead of dignity. This has been attributed to their Western cultural background, which assigns decision making in a clinical situation to the patient and the patient only, whereas the approach may be different elsewhere in the world (but that is a different discussion). Dignity is a more fundamental concept – basic not only to autonomous decision making, but also to, for instance, non-discrimination – and this is what the ICRP document (ICRP 138 2018) is referring to in particular.

Together with justice, which is common to biomedical ethics and ethics of radiological protection, dignity forms the basis of the third principle of radiological protection – dose limitation: '*The total dose to any individual from regulated sources in planned exposure situations... should not exceed the limits specified*'. (ICRP 103 2007) The main purpose here is a fair distribution of risks, and the avoidance of using one person for the purposes of others, which could happen if one individual would be exposed to a relatively high risk in order to save many from a relatively low one (Table 2.1).

TABLE 2.1 Comparison between the *Principles of Biomedical Ethics* (1979) and ICRP Report 138 on *Ethical Foundations of the System of Radiological Protection* (2018)

Principle of Biomedical Ethics (Beauchamp and Childress 1979)	**Core Values of Radiological Protection (ICRP 2018)**
Respect for Autonomy	Dignity
Non-Maleficence	Beneficence and Non-maleficence
Beneficence	
Justice	Justice
	Prudence

2.3 CLASSICAL ETHICAL THEORIES AS A BASIS

In the past, whenever the ethics behind the ICRP system of radiological protection was addressed, it was in the context of European philosophical traditions. Individual authors, among them members of ICRP (Clarke 2003; Streffer et al. 2005; González 2011), have identified arguments primarily from utilitarian and deontological ethics, sometimes referred to as virtue ethics and other theories as well.

Utilitarianism has arguably had the stronger impact, at least during the first few decades, of the ICRP. It is a concept developed by the British philosophers Jeremy Bentham (1748–1832) and John Stuart Mill (1806–1873). Both considered the outcome, or 'utility', of our actions as the only valid criterion for their moral goodness or badness: If what we do causes more benefit than harm it is good, if it causes more harm than benefit it is bad. This is nicely captured in the phrase 'It is the greatest happiness of the greatest number of people that is the measure of right and wrong'. (Bentham 1776; Mill 1861) The clearest reflection in the ICRP system of this kind of thinking is the principle of justification: 'Any decision that alters the radiation exposure situation should do more good than harm'. When it first appeared (ICRP 26 1977) it was worded differently, but equally utilitarian: 'No practice shall be adopted unless its introduction produces a positive net benefit'. The second principle of radiological protection, optimisation, is also based on a consideration of outcomes: 'The likelihood of exposure, the number of people exposed and the magnitude of their individual doses shall be kept as low as reasonably achievable, taking into account economic and societal factors'. It was introduced by ICRP quite early, though at the time it was worded a little differently: 'as low as practicable' (ICRP 1 1959), or 'as low as readily achievable' (ICRP 9 1966). Either way, it became known as the ALARA principle. It is generally understood as urging not only a net benefit, but a maximum of good over harm. Later, ICRP explicitly recommended cost–benefit analysis as a tool for optimization, strengthening the notion that the underlying concept was utilitarian (ICRP 22 1973).

The second influence, deontological ethics, considers as morally valid nothing else than our 'duty' (Greek *deon*) and thus insists that we should never, even if we expected our action to cause more good than harm, neglect the respect for the individual person. And thus, according to the German philosopher Immanuel Kant (1724–1804), we should act in accordance to the Categorical Imperative, which in one of its formulations says, 'Act in such a way that you treat humanity, whether in your own person or in the person of any other, never merely as a means to an end' (Kant 1785). It seems that during the 1970s, the International Commission on Radiological Protection recognized that focusing only on the principle of 'as low as reasonably achievable' did not offer enough protection for the individual. If the 'reasonable' is judged on the basis of a cost–benefit analysis only, we cannot rule out the possibility that somebody would be treated as a means for somebody else's ends. For instance, we might find it reasonable, or even essential, to expose one individual to a relatively high risk in order to save many others from a relatively low one, so that the collective risk can be kept at a minimum. But that would be unfair to the one highly exposed person. ICRP therefore introduced dose limitation as a third principle of radiation protection (ICRP 26 1977): 'The total dose to any individual from regulated sources in planned exposure situations...should not exceed the limits specified'. The recommended dose limits were supposed to keep the risk for professionally exposed radiation workers in a similar range to the occupational risk in other industrial sectors, preferably those that had been classified as relatively safe. Apart from dose limitation, the influence of deontological ethics on radiation protection has been slow to gain ground, but is now the subject of much discussion.

Virtue ethics can be considered as the oldest form of moral theory, going back mainly to Aristotle (384–322 BC). He argued that man is a 'political animal' (Greek *zoon politikon*) (i.e. humans by nature tend to live in a political community). Their well being depends on their functioning well within that social context, which in turn can only be achieved if every individual exhibits certain 'virtues' conducive to the functioning of the whole (Aristotle 350 BCE). Aristotle's ethical thinking is therefore not governed by one overarching principle such as 'utility' or 'duty', but by a whole set of behavioural standards, such as temperance, courage, and generosity, which provide a general orientation on what the right course of action may be. A modern version of virtue ethics is communitarianism, which again emphasizes the social embeddedness of the individual and the need to foster virtues relevant to the community. The first principle of radiological protection – justification – can be read as a general exhortation to contribute to the common good (not so much as a specific instruction to balance profit and harm, which is typically utilitarian), and has therefore been considered as a reflection of virtue ethics in the system developed by ICRP (Hansson 2007).

The problem with all this is that in moral philosophy, utilitarian and deontological theories (or, for that matter, the virtue ethics approach) are considered to be mutually exclusive because they have different priorities. For the utilitarian, as was mentioned above, all that counts is the 'greatest happiness for the greatest number' (Bentham 1776; Mill 1861), whereas the deontologist will insist that you should 'treat humanity, whether in your own person or in the person of any other, never merely as a means to an end' (Kant 1785). It is not clear how a combination of these theories can work, as we can easily think of situations where one would be incompatible with the other. To base the ethics of radiological protection on these philosophical theories, therefore, does not seem a particularly good idea (for further discussion, see Shrader-Frechette and Persson 1997; Persson and Shrader-Frechette 2001; Clarke 2003; Hansson 2007; Gardiner 2008).

2.4 THE NEED FOR A CROSS-CULTURAL APPROACH

Beyond the role of utilitarian and deontological arguments in the evolution of the ICRP recommendations, however, the question can be raised of whether it is at all appropriate in a globalising world to base the recommendations of an international advisory body mainly on ethical theories developed in Europe during the era of enlightenment. Less than 30% of the world's population lives in Europe and the Americas, but over 50% is in Asia and another 20% is in Africa and the Middle East. Can we really expect the majority of mankind to adopt principles of radiation protection developed in a context largely alien to them?

It is true that population numbers do not reflect the relative use of radioactive materials or radiation around the globe, but this situation is gradually changing. According to the World Nuclear Association, there are currently 436 nuclear power reactors in operation, only 120 of them, or 28%, are in Asia, Africa and the Middle East. However, of the 67 reactors worldwide under construction and the 166 reactors planned, 43 and 104, or 64% and 63%, respectively, will be operating outside Europe and the Americas (World Nuclear Association 2015). As for medical radiology, a statistical survey for the year 2011 showed that on average, 131 computer tomography examinations were performed per 1,000 inhabitants of OECD countries. Figures close to this average were reported for Israel (127), Korea (119), and Turkey (112) (Statista 2015). These examples indicate that radiological procedures with relatively high exposures are not restricted to countries with a Western tradition of thought, and countries in Asia, Africa and the Middle East are catching up.

Meanwhile, global approaches to questions of values and norms are becoming more and more common. A first milestone in this development was certainly the 'Universal Declaration of Human Rights' (United Nations General Assembly 1948). In the second half of the twentieth century and especially around the turn to the twenty-first, a number of other international statements on human rights followed, such as the 'Declaration of the Rights of the Child' (United Nations General Assembly 1959), the 'Declaration on Human Environment' (United Nations Conference on Human Environment 1972), the 'Declaration on Environment and Development' (United Nations Conference on Environment and Development 1992), the 'Universal Declaration on the Human Genome and Human Rights' (UNESCO 1997) and the 'Universal Declaration on Bioethics and Human Rights' (UNESCO 2005).

The idea of human rights (i.e. inalienable rights that belong to every human being) of course goes further back in the history of philosophy. Usually, the Stoic school of philosophy (3rd–6th century B.C.) is considered the first to have developed the thought. Bartolomé de las Casas (early sixteenth century) was nevertheless still ahead of his time when he advocated the universality of human rights, stating 'all peoples of the world are humans…The entire human race is one' (Carozza 2003). The idea gained prominence in the era of enlightenment, mainly with John Locke (1689) arguing that 'by nature' human beings have a right to 'life, liberty, and property'. Immanuel Kant (1795) emphasised the interconnectedness of human rights and human dignity and their fundamental importance for the international context, as 'the community of nations of the earth has now gone so far that a violation of right on one place of the earth is felt in all'.

With the rise of globalisation over the last few decades, philosophers have addressed the need for, and possibility of, global ethics from various points of departure. A few examples may suffice here. Jürgen Habermas (1998) speaks of a 'post-national constellation' in which we find ourselves and claims that 'world citizenship… is already taking shape today in worldwide political communications'. Interested in human flourishing and its global dimension, Amartya Sen (2009) has written extensively about the 'idea of justice', which he shows to be central to various cultures around the world, past and present. One of his close associates, Martha Nussbaum (2004) has identified a number of 'core capabilities' which all individuals in all societies should be entitled to, thus constituting the base of her account of 'global justice'. Kwame Appiah (2006) explores the reasonability of cosmopolitanism, which he defines as 'universality plus difference'. While emphasizing 'respect for diversity of culture', he suggests there is 'universal truth, too, though we are less certain that we have it all already'. Sissela Bok (1995) suggests that 'certain basic values [are] necessary to collective survival' and therefore constitute a 'minimalist set of such values [which] can be recognised across societal and other boundaries'. That does not preclude the existence of 'maximalist' values, usually more culture-specific, nor the possibility that they can 'enrich' the debate, but there is a 'need to pursue the enquiry about which basic values can be shared across cultural boundaries'.

One area in which cross-culturally shared ethical principles, values and norms are actively discussed is interfaith dialogue. One outcome of such activities was the 'Declaration Towards a Global Ethic' signed at the Parliament of the World's Religions 1993 meeting in Chicago by representatives of more than 40 different religious traditions. It proceeded from the assumption that 'There already exist ancient guidelines for human behaviour which are found in the teachings of the religions of the world and which are the condition for a sustainable world order' (Küng and Kuschel 1993). Interfaith declarations on more specific topics such as business ethics and environmental ethics have followed (Webley 1996; Orth 2002).

2.5 CROSS-CULTURAL ETHICS, OR COMMON MORALITY

The set of principles suggested by Beauchamp and Childress (1979), which were briefly referred to above, is the most widely applied framework of biomedical ethics around the world. It was not originally conceived as a cross-cultural kind of ethics, but rather turns out to be compatible with such an approach. In the more recent editions of their book, the authors assume that these principles are rooted in 'common morality', which is 'not relative to cultures or individuals, because it transcends both' (Beauchamp and Childress 2013).

Beauchamp and Childress are not really interested in the question of where and how the 'common morality' can be found. When they introduced the term, they just claimed that 'all morally serious persons' (Beauchamp and Childress 1994), or in the latest edition 'all persons committed to morality' (Beauchamp and Childress 2013) would agree with their four principles. This does not seem really convincing to us. More effort is needed to show the principles' cross-cultural validity. Or, should it turn out that they are not globally recognised, to find others that enjoy as wide an acceptance as possible.

One might, of course, think of using empirical research to test the assumption that we have the underlying principles accurate, but we are not convinced that anthropological or cultural studies alone would be meaningful. A universal 'opinion poll' which would find out what people around the globe are thinking about the pertinent questions would just reflect current dispositions and would be very much subject to fluctuations. We have to look for something with greater long-term validity.

Orientation has been provided throughout the ages by the religious and philosophical traditions of the different cultures. Although our Western society is largely secularised, and fundamentalism, fanaticism and extremism have brought religion into disrepute, we cannot ignore the fact that these traditions continue to be of great influence for people not versed in Western secular philosophy. And even in the West, the importance of Christianity is probably still much greater than the number of people attending Sunday church service would suggest. The views of Europeans and Americans have been shaped at least as much by Christian values passed on from generation to generation for centuries, as by the philosophical traditions of the enlightenment era. An analysis of 'common morality' must, therefore, take account of these influences and it would not be wise to neglect them, even in the twenty-first century.

It has therefore been suggested that the most important documents for establishing a 'common morality' are the sacred scriptures of the world's great religions, such as the Vedas and the *Bhagavad Gita* for the Hindus, the sermons of Buddha for the Buddhists, the Torah for the Jews, the New Testament gospels for the Christians, the Quran for the Muslims, the writings of Bahá'u'lláh for the Baha'is, and so on. They provide a framework of orientation for the believers (even though there may be some disagreement regarding their exact meaning), because they are considered to be divinely inspired. Non-believers may of course have some difficulty with this notion, but may at least appreciate that these scriptures reflect values deeply rooted in the various cultures. Another category of useful documents for our purpose is those produced by way of intra- and inter-religious dialogue (see previous discussion), because they already reflect a certain cross-cultural agreement.

There are also relevant cultural expressions outside the context of (organised) religion. Thus, we should not ignore oral traditions in the form of proverbs, stories, legends and myths, especially those of indigenous peoples who have no written records. We should also take into consideration secular texts of various kinds that have had a formative influence over the centuries. The Hippocratic Oath and its contemporary forms come to mind, or the works of certain philosophers of ancient Greece and China (even if Confucius' writings are perhaps more appropriately classified as sacred scripture). In addition to these time-honoured traditions, some modern documents like the 'Universal Declaration of Human Rights' or the 'Universal Declaration on Bioethics and Human Rights' (again, see previous discussion) have been suggested to already constitute 'common heritage of humankind' (ten Have and Gordijn 2013).

2.6 THE VALUES

Below, we are presenting arguments that the four principles of biomedical ethics are rooted in those written and oral traditions of mankind. Similarly, the ICRP report on 'Ethical Foundations of Radiological Protection' in its Appendix C provides evidence for the validity across different cultures of the four core values and three procedural values suggested. Those concerned with the ethical aspects of radiation protection in medicine come from a broad spectrum of medical and allied professions and will in all likelihood be more familiar with systems of medical ethics, including Beauchamp and Childress' work, than with the relatively recent ICRP proposal. We will discuss the four principles of biomedical ethics in their original form first, and then additionally one core value and one procedural value from the ethics of radiological protection. This will provide the Pragmatic Value Set that will be applied in the next two chapters.

2.6.1 Respect for Autonomy

When it comes to cross-cultural validity, the first of the four principles of Beauchamp and Childress is probably the most problematic. It has been criticised as being 'more or less ethno-ethics of American society' (Fox 1990; Matsuoka 2007), but of little relevance elsewhere in the world. In particular, some authors claim that people of Asian background would generally not agree with it, or at least define it differently from Beauchamp and Childress (Fan 1997; Fagan 2004; Kimura 2014). In 'Principles of Biomedical Ethics' the role of this principle is to ensure that the patient is the main decision maker in his or her own case (Beauchamp and Childress 1979). In 2017, it was made part of the 'Physician's Pledge' of the World Medical Association, a modern successor to the Hippocratic Oath (whereas earlier versions had not referred to it explicitly) (World Medical Association 2018). An important corollary of this principle is the concept of 'informed consent', which means that neither therapy nor research can be carried out without the agreement of a competent patient. This understanding of autonomy is certainly common in what we call the West, but not necessarily in other parts of the world. There is at least anecdotal evidence that in Latin America, Muslim countries, Africa, China and South East Asia, the decision making is not primarily a privilege of the individual patient, but is also very much a matter of the patient's family (Justo and Villarreal 2003). Further, it does not appear as if that was to be considered just a current phenomenon, whereas the written and oral traditions actually placed emphasis on autonomy as it is now understood in the West. Nevertheless, there are quite a few Christian (Clarfield et al. 2003; Reilly 2006), Muslim (Aksoy and Elmali 2002; Rathor et al. 2013) and Confucian (Nie, quoted in Justo and Villarreal 2003; Tsai 2005) authors who assert that the principle is fully compatible with their world view. Others disagree, although they usually do not go as far as suggesting that respect for autonomy has no validity. They are just concerned about its relative importance vis-à-vis other principles. We will come back to this question of balancing different moral claims later.

2.6.2 Non-Maleficence and Beneficence

'To abstain from doing harm' is one of the central features of the Hippocratic Oath (Edelstein 1943), which was later adopted by Jewish, Christian and Muslim physicians (Pelligrino 2008). The principle is also mentioned, albeit indirectly, in similar texts from ancient China (Tsai 1999). Of course, it has always been understood that sometimes pain has to be inflicted to achieve healing and thus non-maleficence has to be weighed against beneficence. To work 'for the good of the patient' is part of the Hippocratic Oath as well, and it also features quite prominently in the mentioned Chinese medical texts.

More generally (i.e. outside the context of medicine) both beneficence and non-maleficence can be seen as core principles in any system of religious ethics. A central concept of both Hinduism and Buddhism is *ahimsa*, which means kindness and non-violence to all living beings. The *Bhagavad Gita* praises the 'gift which is made to one from whom no return is expected', whereas the *Dhammapada* states, 'A man is not great because he is a warrior or kills other men, but because he hurts not any living being'. Both the Torah and the Gospel express the same thought in a different way by exhorting everybody to 'love your neighbour as yourself'. More concretely, the Talmud observes that 'to save one life is tantamount to saving a whole world', while the apostle Paul suggests that 'whenever we have the opportunity, let's practice doing good to everyone'. The Quran asserts 'Whoever rallies to a good cause shall have a share in its blessings; and whoever rallies to an evil cause shall be answerable for his part in it'. Nevertheless Islamic jurisprudence has the guideline that 'if a less substantial instance of harm and an outweighing benefit is in conflict, the harm is forgiven for the sake of the benefit' (references in Zölzer 2013).

2.6.3 Justice

The 'Golden Rule' is one of the most common ethical guidelines around the world. It is found in every single tradition one may choose to look at, and even its wording is strikingly uniform. A few examples must suffice: 'One should never do that to another which one regards as injurious to one's own self'. (Hindu) 'Hurt not others in ways that you yourself would find hurtful'. (Buddhist) 'Never impose on others what you would not choose for yourself'. (Confucian) 'That which is hateful to you, do not do to your fellow. That is the whole Torah; the rest is the explanation; go and learn'. (Jewish) 'Therefore whatever you want people to do for you, do the same for them, because this summarises the Law and the Prophets'. (Christian) 'None of you [truly] believes until he wishes for his brother what he wishes for himself'. (Muslim) 'If thine eyes be turned towards justice, choose thou for thy neighbour that which thou choosest for thyself'. (Bahá'í) Because of its general acceptance, this rule is also foundational to the above-mentioned 'Declaration Toward a Global Ethic' of the Parliament of the World's Religions 1993. It is obvious at least from some of the versions quoted here that the Golden Rule can also serve as support for the principles of non-maleficence and beneficence. But, it seems to us that its greatest importance is for the idea of justice. It asks everyone to consider the interests of the other as if they were his or her own, and thus demands reciprocity (references in Zölzer 2013).

Justice as such is verifiably an element of 'common morality' as well. The *Bhagavad Gita* contains the promise that 'He who is equal-minded among friends, companions and foes… among saints and sinners, he excels'. In the Sermons of Buddha, a similar statement is found: 'He, whose intentions are righteousness and justice, will meet with no failure'. The Psalms observe that, 'He loves righteousness and justice; the world is filled with the gracious love of the Lord', whereas in the introduction to the Proverbs the reader is assured that here he will acquire 'the discipline that produces wise behaviour, righteousness, justice, and upright living'. Muhammad advises his followers to be 'ever steadfast in upholding equity…, even though it be against your own selves or your parents and kinsfolk'. And Bahá'u'lláh writes that 'No light can compare with the light of justice. The establishment of order in the world and the tranquillity of the nations depend upon it' (references in Zölzer 2013).

A look at secular philosophy will be instructive here, as justice has not only been of prime importance since Antiquity, but has also been systematically studied early on (Johnston 2011). Aristotle, for instance, distinguished between different forms of justice, and his analysis has exerted decisive influence on later thought. The form that seems to be implied by the sacred scriptures quoted above is 'distributive justice'. It concerns the allocation of goods and burdens, of rights and duties in a society. But even this one form can be viewed from different perspectives. Which allocation of goods and burdens is just? An egalitarian one, one that considers merits, one that considers needs, or one that respects historical developments? We mention these possibilities here in order to indicate that the popular identification of justice with equality is not generally correct. It may be wrongly encouraged by the word's resemblance with equity, which however refers to a fair, or just, distribution of goods and burdens, not necessarily to an equal sharing. Aristotle's recommendation is to 'treat equals equally and unequals unequally'. But how exactly to do that has been a matter of centuries-long philosophical debate. It also needs to be a topic in cross-cultural discourse. This question remains unresolved for now.

2.6.4 Prudence

As mentioned above, in its report on 'Ethics of Radiological Protection', ICRP identifies four core values that are thought to have been fundamental for the evolution of the system of radiological protection established by the Commission. Whereas three of the core values are very similar to the principles of Beauchamp and Childress (replacing Respect for Autonomy by Dignity – see also Chapter 6 – and combining Non-maleficence and Beneficence as one value), the fourth is new: Prudence.

In recent decades, there has been much discussion about the 'precautionary principle', especially in the context of environmental issues. For instance, the United Nations Conference on Environment and Development in Rio de Janeiro 1992, also called the Earth Summit, proposed the following: 'Where there are threats of serious or irreversible damage, lack of full scientific certainty shall not be used as a reason for postponing cost-effective measures to prevent environmental degradation'. (United Nations Conference on Environment and Development 1992) Another important version is the one drawn up by a group of scientists from different disciplines gathered at the Wingspread Conference 1998: 'When an activity raises threats of harm to human health or the environment, precautionary measures should be taken even if some cause and effect relationships are not fully established scientifically' (Wingspread Conference 1998).

Of course, the principle in its modern form cannot be expected to appear in the written and oral traditions of different cultures. Exhortations to prudence, however, are ubiquitous, and are generally interpreted by people referring to those traditions for orientation, as suggesting a precautionary approach. Thus, in the *Mahabharata*, Krishna advises to 'Act like a person in fear before the cause of fear actually presents itself', whereas Shotoku Taishi, the first Buddhist regent of Japan, puts it this way: 'When big things are at stake, the danger of the error is great. Therefore, many should discuss and clarify the matter together, so the correct way may be found'. Confucius simply says 'The cautious seldom err'. In Proverbs, we find the following statement: 'Those who are prudent see danger and take refuge, but the naïve continue on and suffer the consequences', and Muhammad reportedly counselled one of his followers who complained that God had let his camel escape: 'Tie up your camel first, then put your trust in God'. For an explicit reference to the Precautionary Principle I will give just one example, the statement of a representative of the Australian Aboriginals and Torres Strait Islanders: 'Over the past 60,000 years we, the indigenous people of the world, have successfully managed our natural environment to provide for our cultural and physical needs. We have no need to study the non-indigenous concepts of the precautionary principle [and others]. For us, they are already incorporated within our traditions' (references in Zölzer 2013).

2.6.5 Transparency

Of the three procedural values identified by ICRP in their report on 'Ethics of Radiological Protection', we will here only discuss transparency in some detail, because it seems to us that it has the greatest practical importance for medical radiology. Accountability and inclusiveness will be addressed later (Chapter 6).

In the literature about risk communication in general (not necessarily related to radiological protection) a number of different terms have been suggested, which emphasize different aspects of the concept: honesty, truthfulness, veracity, trustworthiness, also frankness, candor and openness. Lambert (1999) prefers 'honesty', because not only should the communicator transmit truth, but also should he or she be clear about information gaps: '... it is unethical to not communicate the uncertainty in knowledge and misrepresent one's perception of the risk (as an objective assessment)'. For Covello (2003), 'truthfulness' is the central concept, which leads him to recommend: 'If in doubt, lean toward sharing more information, not less' and 'Discuss data and information uncertainties, strengths and weaknesses'. Beauchamp and Childress (1979) see 'veracity' as one of the guiding principle for communication in the biomedical context, and see it as a close correlate to 'respect for autonomy'.

Honesty, truthfulness, veracity and trustworthiness are unquestionably virtues that have their place in any religious and philosophical tradition. In the *Mahabharata* we find that 'it is always proper to speak the truth', and Buddha describes his true follower as a 'straightforward person... open and honest'. Confucius states: 'Every day I examine myself... In intercourse with my friends, have I always been true to my word?' Similarly, in the Book of Job, the main character declares: 'My lips will not speak falsehood, and my tongue will not utter deceit'. The Gospel of Matthew contains the following exhortation: 'But let your communication be, Yea, yea; Nay, nay: for whatsoever is more than these cometh of evil'. The same terseness is found in the *Qurán*: 'Have fear of God, and be among the truthful'. And the Bahá'í writings contain this observation: 'Truthfulness is the foundation of all human virtues. Without truthfulness, progress and success are impossible for any soul' (references in Zölzer 2016).

2.7 THE IMPORTANCE OF BALANCING

From the foregoing, it seems clear that the system of principles developed by the International Commission on Radiological Protection is indeed based on values that are shared across cultures. They can be traced back to the religious and philosophical traditions that have provided moral guidance for people around the world over the centuries. That is not to say that secular ethics is wrong and useless, but just that a degree of worldwide consensus already exists and is reflected in those traditions. It is also apparent that the values discussed above are similar, if not identical, to the four principles of biomedical ethics suggested by Beauchamp and Childress, which the authors consider to be rooted in 'common morality'. Cross-cultural validity can be demonstrated both for the core values of the radiation protection system (beneficence/non-maleficence, prudence, justice, human dignity) and for the procedural values which are to guide its implementation (transparency, accountability, inclusiveness – the latter two being further discussed in Chapter 6). Whether radiation protection in practice has always and everywhere reflected these values is a different question, but there is certainly a growing awareness of their importance (as has already been discussed in Chapter 1 and will be more elaborated in Chapter 3).

One aspect needs to be emphasised in this context: The values discussed above, similar to the principles of biomedical ethics, have only *prima facie* validity, which means they apply as long as there is no conflict between them. If there is, they need 'balancing' (i.e. their relative importance must be weighed). It is probably here where cultural specificity can come into play. Beneficence and human dignity, to give just one example, are held in high esteem everywhere around the world, but it is not always possible to implement both of them to the same extent. And if a conflict arises, not everybody everywhere will give the same answer to the question which of the two is to prevail. Consider, for instance, a case where the patient is reluctant to have an X-ray examination, although the physician finds it difficult to decide on an adequate therapy without radiological evidence. Should a paternalistic approach be taken, downplaying the radiation risks and imposing the examination, because it is in the patient's best interest? Or should the approach be patient-centred, leaving the decision to the person affected, although it is clear that the risk of applying the wrong therapy is much greater than any radiation risk? The answer to this question may be different in Korea, the United States, and the Czech Republic, depending on whether beneficence or respect for autonomy takes precedence. Some degree of plurality is certainly acceptable, or even desirable, but we need be aware of the differences and discuss whether we want to retain them or rather develop a common approach. Making the values relevant to radiological protection explicit, and assessing their cross-cultural validity, will help us in this endeavour.

2.8 THE 'PRAGMATIC VALUE SET'

Before the ICRP report on 'Ethical Foundations of the System of Radiological Protection' was published, a number of authors had addressed questions of ethics in the context of radiology and radiotherapy. Our own contribution was the proposal of a 'pragmatic value set' that would support practitioners in their decision making. We assume that most of those working in medicine and the allied professions will have some familiarity with the fundamentals of medical ethics, and hence feel it is reasonable to take Beauchamp and Childress' set of principles as a point of departure.

TABLE 2.2 Comparison between the Core Beauchamp and Childress/the ICRP Values, as Presented in Table 2.1, and the Proposed 'Pragmatic Value Set'

Principle of Biomedical Ethics (Beauchamp and Childress 1979)	**Core Values of Radiological Protection (ICRP 2018)**	**Pragmatic Value Set (Malone and Zölzer 2016)**
Respect for autonomy	Dignity	Dignity/Autonomy
Non-maleficence Beneficence	Beneficence and Non-maleficence	Non-maleficence/ Beneficence
Justice	Justice	Justice
	Prudence	Prudence/Precaution
(Veracity)	(Transparency)	Honesty/Transparency

Note: (Veracity) and (Transparency) in brackets are considered to belong not to the basic Beauchamp and Childress (1979), or ICRP (2018) systems, but represent procedural principles or values.

We added two principles, or values, that we thought were of particular importance for medical applications of radiation, precaution and honesty. These six (or five, as we put beneficence and non-maleficence together) very much overlap with the set suggested by ICRP, although the wording is different in some instances. Let us go through them one by one again (see Table 2.2).

The first principle of Beauchamp and Childress is respect for autonomy, and we definitely agree that it is fundamental for medical radiology as for any other area of medicine. This is further reinforced in the 2017 revision of the Geneva Declaration (World Medical Association 2018), which will be further discussed in Chapter 3. ICRP preferred to list dignity as their (fourth) core value. As mentioned above this seems to be the more basic concept, with respect for autonomy following from it, but it also has other aspects which are perhaps secondary for physician-patient relations. ICRP may have preferred dignity over respect for autonomy because of the necessity to cover a wider spectrum of situations, among them, for instance, nuclear accidents and their aftermath. When it comes to the possibility of continued living in contaminated areas or evacuation from them, these other aspects of dignity, briefly speaking human rights considerations, may play a bigger role. That they are not irrelevant for medical radiology either will be discussed in Chapter 6. Here we note that our proposal for the first 'pragmatic value' was the combination 'autonomy/dignity', trying to equally embrace both concepts.

Beauchamp and Childress list beneficence and non-maleficence as two separate principles. This makes sense, as just looking at the net profit of a radiation application can solve not every question. In many cases, it is necessary to look at the harm itself, or at non-maleficence without referring to beneficence at all. This will be obvious from some of the scenarios presented in Chapters 4 and 5. Nevertheless, in order to make it 'pragmatic', we put 'non-maleficence/beneficence' on the second position. This is also what ICRP has done.

Justice is the fourth principle of biomedical ethics and the third core value of radiological protection. In our proposal it also occupies the third place.

Our original idea for the fourth principle was precaution, because of the very extensive and wide-ranging discussion about the precautionary principle mentioned above. We recognised, however, that first drafts of the ICRP report on ethics (as well as its final version) preferred prudence. This was probably to some extent due to the fact that the precautionary principle has sometimes been abused or applied overzealously. Prudence certainly contains precaution, but it also takes in other factors, which is what the optimisation principle of radiological protection does when it demands consideration of economic and societal factors. As both terms seem to have their advantages and disadvantages, we put on the fourth position of our pragmatic set the combination 'prudence/precaution'.

Finally, we thought of honesty as the most important procedural value for medicine. Again, it was clear that ICRP would rather settle for transparency, which encompasses honesty, but also gives a hint to those communicating about radiation risks that, in addition to speaking the truth, they should speak it in a way that is comprehensible and open to the scrutiny of those affected. As our fifth and last value we therefore chose 'honesty/transparency'.

In the next chapters, we will see how these five (or maybe seven, depending on how you count; see Table 6.1) can be made to work together to solve practical questions of radiological protection in medicine.

CHAPTER 3

The Pragmatic Value Set

Contexts and Application to Radiation Protection in Medicine

3.1 INTRODUCTION

As mentioned in Chapter 1, both medical imaging and therapeutic applications of radiation continue to increase, as do the doses involved. Many clinical procedures, practices and equipment types, that are now commonplace, did not exist in the 1980s. Yet, the ethical basis for these practices has not seen a corresponding level of engagement. Imaging applications are most frequently found in radiology and nuclear medicine. Therapeutic interventions mainly occur, though not exclusively, in radiation oncology. Exceptions include use of radiological imaging in support of other therapies, for example, the placement of cardiac stents, treatment of functional diseases including thyroid problems, or localisation of injection sites in pain clinics. Thus, the benefits of radiology and radiotherapy are not in doubt as they have enhanced the effectiveness of medical practice (ICRP 2007b; WHO 2018b). In addition, both have successfully hosted a technology transfer into healthcare on an exceptional scale.

Nevertheless, imaging patients can, and do, receive significant exposures (NCRP 2009; EC 2015b), often without receiving any information and, sometimes without commensurate benefit. Likewise, there is no escape from the fact that radiation dosage in radiotherapy was, and will, continue to be a major issue. The intent in radiotherapy is usually to deliver an adequate tumoricidal dose to the targeted tissue while preserving the health and function of the adjacent tissues. These somewhat contradictory aspirations require great insight into the surrounding medical, radiobiological, dosimetric and ethical considerations, and in practice present notably different contexts to those prevailing in imaging. The pragmatic ethical framework for both, as outlined in Chapter 2 and further considered here, can serve both well and is illustrated in action in the scenarios in Chapters 4 and 5.

The first half of this chapter extends Chapter 1 in surveying some of the contexts for the pragmatic system. These include aspects of professional and clinical practice, including questions on uncertainties in radiation risk in both imaging and radiotherapy as detailed in the International Commission on Radiological Protection (ICRP) recommendations, in the regulatory framework, and in the radiation protection of patients in practice. The approach taken is, in part, exploratory and tentative and will need further work to fully integrate it into practice of radiation protection in medicine. In the second half of this chapter, each of the five values in the pragmatic set from Chapter 2 is individually explored, with respect to specific issues pertinent to their application in radiology or radiotherapy. There is, inevitably, overlap between both halves, which we have endeavoured to minimise.

3.2 SCENE SETTING

3.2.1 Uncertainty, Communication, Risk and Sceptical Doctors

In Chapter 1, we noted that the dose and probable risk associated with some procedures, for example, traditional multislice cardiac CT examinations, could be significant. The doses involved may be 10–20 milliSievert (mSv) or even more, depending on the type of scan, the age of the patient, and the scanner age and settings. The associated probable cancer risk has been estimated to be as large as 1 in 1,000 per scan. This may be increased by factors of up to 2–10 for young women, girls, or children (Amis et al. 2007; NCRP 2009; NCI 2012; EC 2015a, 2015b; Brenner and Hall 2007). There is a real problem in the health professions on how to deal with this possibility. There is both a lack of conviction and a lack of good debate about probable cancer incidence following exposures at the higher end of the dose spectrum in diagnostic imaging (Malone et al. 2012, 2015). This is compounded by failure of the science communities to find effective, publicly acceptable and transparent ways of communicating about dose and risk to health professionals, and to patients. Part of the problem is that at the (relatively) low doses for diagnostic examinations, there is a dearth of direct evidence. Nevertheless, good, intensively studied data is available from the A-bomb survivors and is the most used epidemiological source for the relationship between attributable cancer risk and radiation dose. Estimates for radiology are derived by extrapolating the dose effect curve from higher doses. However, a relationship is now convincingly seen down to about 100 mSv and possibly lower (i.e. dose levels overlapping the range of a few CT scans) (Horton 2011; Shah et al. 2012).

Several models of the dose response for cancer incidence after irradiation are available. While it is not possible to select between them, the most authoritative voices available conclude that a Linear No Threshold (LNT) model remains a valid conservative choice for calculating risks at low doses (BEIR VII 2006; ICRP 2007a; NCRP 2018). Some workers treat the LNT conclusions as though they provide definitively established risk levels which do not need to be qualified by any sense of uncertainty. The LNT model is important to the ICRP in reaching its recommendations, though it may be qualified to compensate for special features of medical exposures dose rate. Nevertheless, it has become an important part of the background to the development of law and regulation on radiation protection in most countries (BEIR VII 2006; ICRP 2007a, 2007b). Hence, it is central to the system we now have and cannot be lightly dismissed without offering an alternative approach that would also be able to serve this purpose.

Radiotherapy has joined the uncertain space of low-dose effects with the increased use of IMRT (intensity-modulated techniques). The impact of low-dose to large volumes of uninvolved tissue has created a heated debate on the benefit–risk front. The combination of improved diagnostics and improved treatments has greatly improved survival rates, but this inevitably raises concerns of secondary malignancies or late cardiovascular toxicities at the treatment sites (Carver et al. 2007; NCRP 2011; Travis et al. 2012).

Many radiologists, cardiologists, medical physicists and others are radiation damage sceptics, and move seamlessly from the view that there is no definitive evidence of low-dose damage, to the position that there is no damage and behave accordingly. As they are sceptical about risk, they tend toward discarding consideration of it, although the basis for such a position is a *non-sequitur*. In this context, a now withdrawn statement from a well regarded professional organisation is worth mentioning. It is helpful in understanding the position leaders in the professions adopt (Brenner and Hall 2007; Hendee and O'Connor 2012; Malone et al. 2015). Here is an extract:

> Risks of medical imaging at effective doses below 50 mSv for single procedures or 100 mSv for multiple procedures over short time periods are too low to be detectable and may be non-existent. Predictions of hypothetical cancer incidence and deaths in patient populations exposed to such low doses are highly speculative and should be discouraged... These predictions are harmful because they lead to sensationalistic articles in the public media that cause some patients and parents to refuse medical imaging procedures, placing them at substantial risk by not receiving the clinical benefits of the prescribed procedures.

From the perspective of the public, it is probable that the statement would be viewed as, at best, paternalistic without provision of information the patient might use in reaching a decision on his/her own behalf.

While the statement is withdrawn, similar comment continues to be available (Hendee and O'Connor 2012). On the other hand, Shah et al. have taken the above statement as a counter example to both the precautionary principle and to justification (i.e. benefits are emphasised without reference to risks) (Shah et al. 2012). It would be easy to read the statement, and some of its predecessors, as dismissing the risks for most radiology, with little regard for the scientific position that there may be some risk (Shah et al. 2012; Malone et al. 2015). A more extreme version of this is encountered among some practitioners who are LNT sceptics, and do not advise patients about risk and omit it in their approach to diagnosis or treatment.

Low-dose scepticism is also found in radiotherapy, for example, when IMRT was introduced, the short-term gains in effectiveness were welcomed without much reference to increased dose to non-target organs or the associated longer-term risks. IMRT techniques in adult treatments of oropharyngeal, gastrointestinal, and genitourinary disease have grown exponentially with little restriction. However, most centres are cautious with its paediatric application, something that is neglected in some adults (Rembielak and Woo 2005).

The ACR White Paper notes that some physicians are very knowledgeable on these issues and incorporate such information into their decisions, but others do not routinely do so (Amis et al. 2007; Chapter 1). Radiation protection must, in practice, find its place between the extremes of unjustified fear and unconcerned use of radiation. The message should be that there may or may not be a risk; we don't know for typical diagnostic or out-of-field radiotherapy doses. But we do know a lot but are being cautious (i.e. using the precautionary principle). The most studied and critically evaluated position of the scientific community for future cancers/deaths from scans is the LNT hypothesis (NCRP 2018). This should be used in communication with staff and/or patients and should be qualified with accessible explanations of the uncertainties involved. The diverging attitudes and approaches that now characterise the professions and the public harbour the possibility for serious misunderstandings and/or conflict.

3.2.2 ICRP Recommendations and Medical Exposures

The system of radiation protection in the great majority of countries in the world is based on the recommendations of the ICRP. The publications of the ICRP are purposely built for radiation protection, and are based on a solid scientific evidential base, combined with value judgments and experience, to allow it be applied to practical problems in industry, medicine, education, research and in everyday life. The values on which the ICRP relies were until recently, as already mentioned, implied rather than explicit. The source documents in which ICRP principles are most clearly articulated are the recommendations of the main commission in publications 26, 60 and 103. In addition, ICRP publication 138 (released in early 2018 as this book was being finalised) identifies the ethical basis for the system of radiation protection as a whole (ICRP 1977, 1991, 2007a, 2018). With respect to medical uses, publication 105 is also important, though it adds little to the principles (ICRP 2007b). However, ICRP 105 identifies unique aspects of the use of radiation in medicine, including that, in the case of patients, the exposures are deliberate, voluntarily accepted and consented to, and that no dose limit applies to them. In addition, an ICRP publication dealing with the ethical issues specific to medicine is awaited.

The system for regulation of the use of radiation in medicine follows the general recommendations of ICRP as expressed in standards and directives by the IAEA – a member of the UN family – the European Commission (EC) and other regional/national bodies (EC 1997, 2013; IAEA 2014). These requirements are transposed into national law in many countries.

The principles of *justification*, *optimisation* and *dose limitation* are central to the ICRP system:

- Justification: of the activity
- Optimisation: using a dose as low as reasonably achievable (ALARA)
- Dose limitation: application of dose limits and dose limitation strategies; dose limits do not apply to patient exposures

The relationship between the ICRP principles and the principles/values of medical ethics is, at least on the surface, far from self-evident and requires much exploration (Chapters 2, 6 and 7). There are also likely to be important differences in the way values are currently deployed for medicine/dentistry, on one hand, and radiation protection in general, on the other. This suggests that as well as areas of close overlap, there could well be areas of conflict in the way the values are expected and used in different fields.

3.2.2.1 Justification

Key to effective implementation of an ethical framework in radiological imaging and radiotherapy is ensuring that those referred for procedures need them. Justification is a means of ensuring this. It requires that medical exposures must benefit the patient. In practice, the processes it advises be used are based on benefit–risk analysis and most discussion in the area is limited to this issue. ICRP identifies three levels of justification for all medical exposures:

Level 1: Justification of use of radiation in medicine. At a general level, the use of radiation in medicine is accepted as doing more good than harm. Its justification is taken for granted. While this is likely to continue from a legal perspective, it may be challenged ethically from several points of view. For example, there are concerns about the overall legal framing of the system of radiation protection (Meskens 2016b). Likewise, the relationship between the ICRP principles and ethical values in medicine must be related to the heritage of medical ethics, particularly as articulated in Beauchamp's and Childress' system. Finally, the narrow basis (benefit–risk) on which the operation of the ICRP system has been based to date will be challenged by dignity/autonomy, prudence and justice from the pragmatic set, particularly in respect of justification.

Level 2: Justification of a defined radiological procedure. The second level concerns particular procedures with specified objectives (e.g. chest radiographs for patients showing relevant symptoms). The aim is to ensure the procedure normally improves management of the patient. This is a matter for national professional bodies, health and regulatory authorities.

Level 3: Justification of a procedure for an individual patient. The third level concerns the application of the procedure to an individual patient, which should be judged to do more good than harm to the individual in his/her circumstances. The second and third levels of justification are those that apply in day-to-day medical practice.

Recent formulations of advice and regulation have somewhat diluted this very clear position. For example, it is suggested the total benefit from a medical procedure may include not only the direct health benefits to the patient but also the benefits to the patient's family and to society. At Level 3, this has obviously been viewed as acceptable and provides, for example, the basis for screening programmes involving radiation. In addition, it has been recommended that the risks include the risks to the workers and others involved. However, the precedents created are not without problems, and this blurring of categories needs a careful critique to reposition the benefit–risk considerations involved, and to sustain justification of the exemption from dose limitation for medical exposures (ICRP 2007a, 2007b; EC 2013; IAEA 2014).

The prevailing social environment has raised the level of openness, accountability and transparency expected of professionals and institutions. In larger institutions, imaging departments can be huge enterprises, with several hundred staff undertaking 500–1,000 examinations a day, possibly several hundred thousand per year. This is imaging on an industrial scale and the skills to manage it effectively are not always available. The well being of the individual patient may be lost in such large systems. The funding and referral arrangements in both public and private systems can encourage radiologists to accept inappropriate referrals.

Good practice in radiology relies on a core principle that each examination is justified for the patient involved. There are serious problems with the implementation of justification in practice, with an estimated 20%–50% of radiological examinations being unjustified in practice, and the figure can be as high as 60%–77% in special cases (e.g. lumbar spine examinations, or cardiac angiography). In addition, an international workshop with participation from 40 countries and the relevant international organizations concluded that 'There is a significant and systemic practice of inappropriate examination in radiology' (IAEA 2011; Malone 2011a; Malone et al. 2012).

A joint IAEA/WHO initiative, The Bonn Call to Action, recognises this problem and gave it a high priority (Bonn Call 2016). This arose from a 2012 conference called to set an agenda for radiation protection of the patient for the next decade (IAEA 2015). It proposes it be addressed using a three-prong approach, *The Three A's*, aiming to bring about improvements in *Awareness*, *Appropriateness* and *Audit (Clinical)* as the key tools to be applied in day-to-day practice. This approach has been adopted by numerous international, regional, national, professional and local bodies including, for example the IAEA, WHO, EC, The Heads of European Radiological protection Competent Authorities (HERCA), European Society of Radiology (ESR), The Nordic Radiation Protection Authorities, and many hospitals, among others. Aspects of The Three A's have been included in many programmes including those of the Eurosafe campaign of the European Society of Radiology (ESR), The American College of Radiology (ACR), Image Gently, Image Wisely, Choose Wisely and US financially driven incentives to limit overutilisation (EC 2004; ACR 2013; RCR 2017; ESR 2018; Eurosafe 2018; Image Gently 2018; Image Wisely 2018).

3.2.2.2 Optimisation

Optimisation of protection in medicine is usually applied at two levels: (1) the design and construction of equipment and installations, and (2) protocols and working procedures used to guide practice. The aim of optimisation of protection is to maximise benefit while using the minimum dose required to achieve this objective (i.e. to maximise the net benefit). Doing this is likely to have financial and other resource implications, but it may also have social costs including some risk to staff. Optimisation of protection implies keeping the doses 'As Low As Reasonably Achievable' (ALARA), economic and societal factors being taken into account' (ICRP 2007a, 2007b). However, how to take such factors into account on an objective-evidential basis is fraught ethically, as is the problem of balancing benefits and risks in justification, both of which will be further discussed later in this chapter and in Chapter 6 (Zölzer and Stuck 2018).

The concept of dose constraints is often applied in support of optimisation (or the ALARA principle). For example, in the design of buildings or equipment, the regulator may insist on a maximum dose less than the dose limit, which it is known can be achieved in practice. Dose constraints are not applied to patient doses, but a concept with a somewhat similar intent, Diagnostic Reference Levels (DRLs), is applied. These are further discussed in the section on dose limitation below.

There is considerable scope for dose reductions in diagnostic radiology. Much can be achieved with relatively low-cost measures without loss of diagnostic information, but the extent to which these measures are deployed varies widely. Enormous benefit has derived from observational studies recording the doses for different examination and patient groups, in individual diagnostic rooms, whole departments, and in regional and national surveys (EC 2015b). Work in this area has provided informal benchmarks against which practitioners can evaluate their own performance, and allowed protection for subgroups be optimised (e.g. multislice CT scanning in children). Large spreads, often of one or two orders of magnitude, in the size of the dose per examination are still routinely reported, and while much has been done, there is still much to do in optimising examinations and achieve a good balance of image quality and dose on a global scale.

There is an overlap in the taxonomies of justification and optimisation. For example, if a scan is performed when none is necessary, or if a CT scan is undertaken for a patient when an ultrasound would have been more appropriate, then these are usually regarded as failures of justification. On the other hand, some would also argue that it is also a failure of optimisation as the required imaging outcomes are not achieved at the lowest practical dose. However, in practice matters pertaining to doing the correct necessary examination with the correct modality is generally discussed under the heading of justification. Questions around the exposure level used, the criteria for acceptable equipment functioning, and using a good technique is usually discussed under the heading of optimisation (EC 2012). In practice, enormous progress has been achieved on optimisation, whereas serious work on justification has only emerged in the last decade or so.

In radiation therapy, optimisation includes the need to differentiate between the dose to the target tissue and the dose to other parts of the body. The protection of tissues outside the target volume is an integral part of dose planning and must be optimised provided the dose to the target volume is not compromised. Much recent development in radiotherapy arises from the ambition to deliver treatments tailored to each patient giving adequate target dose while reducing dose to the adjacent structures as much as possible. While a lot has been achieved, there is a lot more to be done. This includes, for example, new and potentially more favourable particle therapies as well as real time treatment planning based on live imaging.

3.2.2.3 Dose Limitation

The concept of dose limits is applied to exposures of workers and the public. However, it is not applied to exposures of patients, so that the discretion necessary in some medical procedures will not be unduly interfered with. Thus, medical exposures are particularly privileged in this regard. The concept of medical exposures is extended beyond patients in the EC's Directives on Radiation Protection and applied to exposures of *Comforters and Carers* to those undergoing medical procedures (e.g. family members) (EC 2013). Thus, such exposures are also exempt from dose limits. However, an advisory system of DRLs is mandated and widely used to discourage excessive dose levels in practice (see below) (ICRP 2007b; EC 2013). Likewise, in radiotherapy, professionally approved advisory guidelines on constraints to organ dose are widely employed (Emami et al. 1991; Bentzen et al. 2010).

3.2.3 Regulation of Medical Exposures and Radiation Protection of Patients

The regulatory framework for radiation protection in general and in medicine relies heavily on the recommendations of ICRP, but the structures and framework for implementation differ. Two organisations have provided exceptional leadership in setting standards and approaches that have been widely adopted. They are the IAEA, a member of the UN family, acting for the international community, and in Europe, the European Commission (EC) acting on behalf of its 28 member states. The IAEA creates non-mandatory standards that often underpin legislation and the framework for good practice internationally (IAEA 2014). Its standards may be enforced when a state is in receipt of aid from one of the IAEA programmes. The EC issues legally binding directives, which member states much transpose into national law, as well as a large body of advisory literature on implementation of its directives (EC 1997, 2013, 2018). There is a high degree of co-ordination between the IAEA and the EC, and in situations involving medical exposures. WHO is also party to these co-ordination efforts and provides advice through its Global Initiative on radiation protection in medicine (WHO 2018).

The EU has a well-developed mandatory framework for radiation protection of patients within its member states. This grew, at least in part, from a perceived lack of compliance with general radiation protection provisions among the medical community, and led in 1984, anecdotally at least, to the first predecessor of its mandatory MED (Medical Exposures Directives), Directives focusing on radiation protection of patients in its member states. The value system informing the MED, and the national regulations deriving from it, is essentially that of the ICRP. All versions of the MED, up to and including the most recent in 2013, and consequent national legislation emphasise the three ICRP principles (EC 2013).

An almost universal requirement of regulations and standards is that those responsible for and administering radiation to patients receive appropriate specialist education and training, and are thereby competent (IAEA 2009; Sia 2010). However, it has been widely demonstrated in an ongoing series of studies that doctors' knowledge in respect of dose and risk, as well as in connection with identification of the most appropriate examination for particular indications, is generally poor. The first study in this area in 2003 demonstrated an alarming deficit unequivocally (Shiralker et al. 2003; IAEA 2011; Malone et al. 2012). Several studies confirming this are now reported every year involving doctors, including radiologists, allied health professions, trainees, students and patients (Shiralkar et al. 2003; Lee et al. 2004, Singh et al. 2015; Campanella et al. 2017). Although there is evidence of improvement since the original report, the deficit in knowledge remains significant.

Within the medical radiation protection communities, the emphasis until the year 2000 was on the safety of workers, other staff, the public and the safe design of equipment and buildings for radiology, nuclear medicine and radiotherapy. Explicit emphasis on radiation protection of patients, particularly in diagnostic imaging, only gained momentum from the year 2000, and an IAEA conference held in Malaga in 2001 was a key catalyst for the international community in the area (IAEA 2001). The National Council on Radiation Protection and Measurements (NCRP), possibly stimulated by Malaga, identified that there was an increase of 570% in individual medical doses in the United States between 1980 and the early years of the new century (NCRP 2009). This arose from both the frequency of examinations and the dose per examination. Closer examination suggested a tsunami of imaging whose epicentre was in CT and nuclear medicine. The individual exposures involved in an abdominal/pelvic CT or some cardiac nuclear scans can be equivalent to several hundred chest X-rays.

Most countries have strict radiation dose limits for the general population (1 mSv per year) and for professionally exposed workers (20 mSv per year).

However, as previously noted, dose limits are not applied to medically exposed patients and a few groups formally defined as medical. Thus, paradoxically, a citizen upon becoming a diagnostic imaging patient loses the protection of a dose limit, and entrusts their care to physicians who often do not know the dose or risk to which they are exposed. Patients can and do receive significant exposures, sometimes larger than the annual dose limit for workers, without receiving information and, in the case of inappropriate examinations, without commensurate benefit. In Europe, this is somewhat offset by the requirement to have a Medical Physics Expert (MPE) to advise and provide support on matters pertaining to radiation exposure of patients (EC 2013).

However, in diagnostic imaging, the absence of a dose limit does not mean there is no attempt to control doses. The concept of DRLs mentioned previously, has been introduced and, by and large, been successfully championed (ICRP 2007b). DRLs are advisory dose levels established for examination types (e.g. a chest X-ray or CT). They may be established for a department, a hospital, regionally or nationally. The DRL is based on a survey of doses in the hospital or region involved. It is set at the third quartile of the dose values observed (i.e. 75% of the examinations surveyed will have received doses less than the DRL). The DRL is an advisory statistical quantity, and not intended to impact individual exposures. However, it is a useful measure of average department, regional or national performance, and if regularly exceeded is a wake-up call to examine practice and see what can be improved.

3.2.3.1 Definitions

Four definitions are provided here for ease of reference (EC 2013; IAEA 2014). For radiation protection purposes, practitioners in diagnostic imaging are separated into those referring the patient and those performing or overseeing the examination, as follows:

Referring practitioner: A health professional who, in accordance with national requirements may refer individuals to a radiological practitioner for medical exposure.

Radiological practitioner: A health professional with specialist education and training in the medical uses of radiation, who is competent to perform independently or to oversee procedures involving medical exposure in a given specialty.

In addition, two types of patient referral or presentation, not traditionally encountered in radiology, now occur:

Self-referral: A physician (e.g. a cardiologist) who has radiological facilities may perform a procedure on a patient instead of referring on to a third party, such as a radiologist.

Self-presentation: A patient may refer themselves for a procedure and directly request a radiology facility to have it undertaken.

These referrals tend to give rise to ethical and financial concerns and increase the use of ionising radiation over and above that which prevails in the traditional approaches involving third-party referrals. They are subject to much comment, and can be of varying standards in terms of reporting and integrating with clinical follow up. Outside formal screening programmes, self-presentation can become so detached from the medical system that the examinations involved may be of greatly reduced value to the self-presenter. Indeed, the term ***presenter***, instead of patient, has been suggested for such situations, although this is challenged (Malone et al. 2016; Schaefer 2018).

3.2.3.2 Radiation Protection of Patients in Medical and Dental Imaging

Radiation protection of patients in diagnostic imaging has moved from being a Cinderella activity to being one whose importance is widely accepted with significant resources being devoted to it. For example, the IAEA has a successful outreach website (IAEA 2018) devoted to it; significant programmes of monitoring patient dose and justification are undertaken in many countries; and numerous professional bodies have become deeply involved in initiatives to promote good practice in the area, a sample of which are cited here (IAEA 2011, 2018; EC 2008, 2009, 2012, 2015a, 2015b; ACR 2013; RCR 2017; ESR 2018; Image Gently 2018; Image Wisely 2018). The objectives of achieving justified examinations at reasonable dose levels preoccupied professionals for many years and were included in the Bonn Call for Action (2016). Identification of the important areas in the Bonn Call drew, at least in part, on ethical sensibilities and values close to those in the pragmatic set. It is unlikely that the objectives of the Bonn Call can be achieved without an explicit ethics framework.

In practice, the culture of radiation protection in medicine has come to rely on professionals avoiding talking to patients about the uncertainties involved and assuring them that everything is fine. This is no longer acceptable, both as a purely practical matter and, more importantly, because of the emphasis placed on the autonomy of the individual and honesty in contemporary thinking. Perhaps this was well captured in the following extract from the *New York Times* (Fazel et al. 2009):

> I think the central driver is more about culture than anything else. People use imaging instead of examining the patient; they use imaging instead of talking to the patient... Patients should be asking the question: 'Do I really need this test? Is the information in this test going to help in the decision-making process?' ... In many cases, there is little evidence that the routine use of scans helps physicians make better decisions, especially in cases where the treatments that follow are also of questionable efficacy.

This reinforces the view that much professional behaviour in the area is effectively ritualised, and part of a culture that is resistant to challenge.

The problems identified in dentistry are like those in medicine, with a few important differences. First, the traditional dental examinations in intra oral radiography were relatively low dose and possibly gave rise to a relatively casual approach to prescription of examinations on the grounds that the very low dose would not do much harm. However, with technological advance, first with relatively routine orthopantomography (OPG) radiography, and more recently cone beam dental examinations, doses have increased, and the dose issues have become more important (PHE/HPA 2010). In consequence, both justification and optimisation have moved centre stage and it is urgent that they be treated seriously.

The most important difference between medicine and dentistry is that the dentist is generally responsible for the prescription of the examination, as well as its execution. In medicine, for example, a GP or gastroenterologist requests an examination, but responsibility for its performance normally passes to another practitioner, a radiologist. In theory, this offers additional protection for justification as both the referrer and the radiologist must be satisfied that it needs to be done. In practice, this protection, in dentistry for example, is missing with the person referring the patient for radiology actually being responsible for, and possibly financially benefiting from its performance. In addition, commitment to good optimised practice in dentistry imposes much greater burdens than those that prevailed prior to the adoption of the newer technologies (Horner 2018).

3.2.3.3 Special Situations

Special ethics concerns arise with all exposures, diagnostic or therapeutic, of sensitive or at-risk subgroups of the population, including women, children, pregnant women and those with possible increased radiosensitivity. In addition, there are classes of deliberate human exposure that may not be regarded as medical. These are further discussed, with examples, in the scenarios in Chapter 4. Medical exposures generally confer benefit on the exposed person, and are conducted under the supervision of medical practitioners with training and authorisation to do so. Borderline situations exist where there is some doubt that all the requirements for *bona fide* medical exposures are met. Examples include: lifestyle radiology, CT screening of asymptomatic people, and unapproved screening programmes. In addition, there is the exposure of volunteers in biomedical research programmes that provide no direct benefit to the volunteers (EC 1998b). With these, dose constraints are applicable to limit inequity, and because there is no further protection in the form of a dose limit (EC 2013; IAEA 2014).

There are also frankly non-medical exposures, such as those undertaken for security, crime prevention, detection of smuggling and those arising from litigation. These are addressed in the European and International Basic Safety Standards (BSSs) (EC 2013; IAEA 2014). They raise important ethics questions and are also addressed in some of the scenarios in Chapters 4 and 5. Advice/guidance is also available on radiation of volunteers in medical research, although it predates recent developments in ethics, and hence would benefit from being reconsidered in light of the pragmatic set.

3.2.4 Radiation Protection of Patients in Radiotherapy

As with radiology, radiation protection in radiotherapy tended to focus on protection of the staff, protection of the public, and on the safe design of equipment and buildings, with little attention to the patient. The IAEA emphasises accident prevention and treatment optimisation when addressing radiation protection of patients in radiotherapy. The ICRP, NCRP, AAPM, ARPANSA (Australian Radiation Protection and Nuclear Safety Agency) and numerous other organisations/professional bodies have safety reports sharing the common thread of protection of the public, the worker, and the patient from accidental irradiation in radiotherapy.

Radiotherapy may, in fact, be among the safest of the treatment modalities available to patients, as it is arguably one of the most critically examined medical disciplines in terms of the functionality and quality of its equipment. This is partially attributable to the level of alarm and the public profile associated with some of the accident reports in the area over the last couple of decades (IAEA 2000). In addition, the ICRP suggested that incidence of accidents will increase if measures for prevention are not taken (ICRP 2000b). The improved safety designs implemented by the manufacturers and standards bodies (IEC 2018), the international beam calibration protocols released by AAPM and IAEA, the release of commissioning guidelines for equipment and treatment planning systems, and quality assurance protocols recommended by professional bodies, have all contributed to a safety culture that characterises most radiotherapy centres today (AAPM 1998, 2009, 2011; NCS 2013; CPQR 2016).

Because of these safety measures, accident rates in radiotherapy are small. Cancer Research UK reported over 14 million new cancer cases worldwide for 2012 (GLOBOCAN 2012). Over 4 million treatment plans were designed for a subset of this population leading to administration of over 50 million treatments during a year. The rate of reported overexposure accidents was about 11 per year, giving an incident rate of about 1 per million patients compared to 1 per 30 patients undergoing surgery, or compared with the incidence of various forms of misadventure once one is hospitalised (Neale et al. 2001; Thomas and Brennan 2001; James 2013; Coeytaux 2015). However, ICRP notes that in many of the accidental exposures, a single cause cannot be identified, and the observed mistreatments are generally due to system failures (ICRP 2000b). A recent AAPM report encourages departments to further examine the implementation of safety measures using failure mode analysis (Huq et al. 2016) to assist in identifying the possible system failures and implement further measures to prevent them. The overlap between risk management and patient protection is prominent in radiotherapy and can perhaps be attributed to the potential severity of damage following accidental overexposure.

However, in this context, it is wise to remember that the patient will not always be damage free when radiation is delivered exactly as intended. The intention and design of radiotherapy is to do biological damage. It must generally be evaluated against a balanced consideration of all five values in the pragmatic set (Chapter 5). Much of the difference between radiotherapy and radiology arises from the deliberate intention to damage tissue in radiotherapy. This is a direct challenge to the beneficence/

non-maleficence value in the pragmatic set, in a way more analogous to surgery than radiology. Unlike the direct damage of a scalpel, sensitivity to radiation damage is dependent on many parameters, such as radiation type, cell type, its level of differentiation and oxygenation, among many others. In addition, technical innovations offer numerous options, such as precision manipulation of the shape and intensity of the beam, and precision targeting of particular cell types with special radionuclide labelled molecules. All these developments are aimed at improving the therapeutic ratio (i.e. the ratio of damage to tumour tissue compared with that to normal healthy tissue). The plethora of treatment options in developed countries, the unprecedented demand for treatment, the possible choices often shrouded in new forms of uncertainty, all demand new and rigorous attention to justification, ethics and values, for the welfare of the individual patient, and particularly to justice in dealing with resource allocation at a social level.

Clearly radiotherapy and the scientific and biologic principles that guide its application have continuously evolved in recent decades. The ethical framework that underlies individual treatment and resource allocation decisions must also be continuously energetically reconsidered and updated. The pragmatic value set, further discussed below, can be applied to these challenges.

3.3 THE PRAGMATIC SET IN MEDICAL/DENTAL PRACTICE

In Chapter 2, the five widely recognised values of the pragmatic set for medical radiation protection, and the basis for them in philosophy and the wisdom literature of the world, were identified. These are essential to complement the ICRP principles and are listed again for convenience in the first column of Table 3.1. The second column contains brief observations on the application of each value as it might occur in medicine.

When referring to justification, optimisation and dose limitation (i.e. the three 'principles' of radiation protection), we continue, as in Chapter 2, to use the term 'principle' and use it exclusively for these ICRP principles. On the other hand, we use the term 'value' when dealing with the pragmatic set, or the additional values discussed in Chapter 6.

At this point it is valuable to note that medical ethics has a tried and tested literature and a global active research community. Even though medical irradiation is generally conducted in clinical facilities, those involved with radiation protection in medicine have seldom looked to medical ethics to contribute to their field. The unconscious assumption

TABLE 3.1 Five Pragmatic Values Underlying the Principles of the ICRP as Applied in Medicine

Number	Value	Comments
1	Dignity and autonomy	Of the individual.
2	Non-maleficence; beneficence	Do no harm; do good.
3	Justice	In the sense of fairness of deployment of resources and risk.
4	Prudence/precaution	Where there is a possibility of serious irreversible harm. As appears in precautionary principle.
5	Honesty	Particularly in being honest and open with patients about their treatments. And being transparent.

Based on: Malone, J.F. and Zölzer, F., *Br. J. Radiol.*, 89, 20150713, 2016; see also Chapter 2.

was possibly that the radiation protection system is complete as presented in ICRP (2007a, 2007b). Clearly, as set out in ICRP 138, the values underlying the radiation protection system must now be taken explicitly on board. The pragmatic value set provides an acceptable approach to this pending an additional report from the ICRP on ethics for radiation protection in medicine (ICRP 2018).

Although advocacy in respect of the values underlying radiation protection in medicine has from time to time been intense, it is somewhat redundant when viewed from within medicine. This is because compliance with a mature system of medical ethics is non-negotiable for medical activities such as radiology and radiotherapy. Working within the pragmatic set fulfils this requirement. In addition, in medicine, the longstanding system of values stretching back to the Hippocratic Oath has been given modern expression in the 2017 revision of the World Medical Association's Geneva Declaration, which is also close to the pragmatic set, particularly in its emphasis on the dignity and autonomy of the individual (Parsa-Parsi 2017). Both the Declaration and the pragmatic set are predicated on the need for care and ethical sensitivity in the way patients are treated and how treatments are delivered. The approach to ethics in Chapter 2 is global, and among other influences, rooted in the medical tradition, as well as being alert to contemporary social expectations (Beauchamp and Childress 2013; Zölzer 2013).

The values from the pragmatic set must be balanced in their application, as their varying requirements will inevitably conflict with each other and with existing practices. The values also need 'specification', (i.e. concrete rules or guidelines must be derived for different areas of application).

For example, dignity and autonomy are essential for justification, and give rise to a requirement for implicit or explicit consent. In addition, as will be discussed below, they come into conflict with non-maleficence (i.e. do no harm). Beauchamp and Childress discuss the practical application of their values and how they may be 'balanced' and 'specified' at length. Their work in these matters is frequently cited and highly regarded. Finally, while the systems of medical ethics and the ICRP system for radiation protection overlap significantly, there is as yet no simple way of mapping one onto the other.

3.3.1 Dignity and Autonomy

Chapter 2 argued that respect for autonomy is based on an understanding of human dignity and that the latter is more easily demonstrable as cross-cultural. In the medical/radiological context, this value protects the position of the patient as the main decision maker in his or her own case. The importance of this and its almost universal acceptability is further emphasised, as detailed above, in the 2017 revised version of the Declaration of Geneva (Parsa-Parsi 2017).

With regard to radiological protection, this implies that the imposition of risk must take account the individual's volition, and this is a prerequisite for the principle of justification. This is, by and large, absent from the justification narrative underpinning both ICRP recommendations and much of national legislation for both diagnostic and therapeutic applications. As envisaged in these sources justification is primarily a matter of balancing the patient benefit from the procedure against its associated risks or harms (ICRP 2007b). Even on this narrow basis, as noted above, the literature suggests that overall 20%–50% of radiological examinations may not be justified (IAEA 2011). Many significant problems arise from this approach to justification, not least the fact that both the benefits and the risks are difficult to quantify in a way that leads to an ethically valid comparison. Recourse to economic factors simplifies matters but can also be ethically problematic. Likewise, it is difficult in practice to account for societal factors (Zölzer and Stuck 2018).

Furthermore, the problem of balancing the probable harm against the benefit is fraught as the available evidence for both often includes much uncertainty, and seldom allows direct comparison of like with like. Usually, when evidence is available, it is comparing apples with oranges, or even with onions. Real evidence of benefits, at the level of improved health outcomes, is often not available. At a less demanding, but more accessible level, evidence on the accuracy of diagnosis and how imaging influences patient management is not as good as it needs to be. Even at the level of basic sciences, the evidence for something that appears simple, for example, the technical measures of image quality and their relationship with diagnostic accuracy, leave a lot to be desired (SENTINEL 2008; EC 2012). Initiatives from the IAEA, WHO, EC, and other bodies emphasise the need to address the question of health outcomes urgently, and this is implicitly reiterated in the Bonn Call for Action which identified improvement in justification as a priority (IAEA 2011; Bonn Call 2016).

This benefit–risk narrative misses the important consent issues which flow from dignity and autonomy of the individual patient (see below). The simplest remedy to this may be to try and achieve agreement to practice justification for radiation protection of patients rigorously within the framework of medical ethics and the Declaration of Geneva, rather than as a freestanding self-sufficient principle judged only on benefit–risk considerations. In addition, as noted above, the benefit–risk analysis takes place in a context of uncertainty about both the benefit and the risk (Chapter 7). Briefly referring to some of the other values, particularly non-maleficence, the precautionary principle and honesty (all treated below), it is reasonable to take the view that patients have the right to be made aware of this, and that physicians have a duty to inform them. WHO places a high value on identification of the right patient, for the right examination, at the right dose. However, it is likely that the practical consequences of embracing the concept of dignity and autonomy, and the other values, will lead to significant developments in how justification is practiced.

The IAEA suggest moving towards affirming the patient's autonomy, and this immediately raises the consent issue. In practice, consent for radiological examinations is not generally sought, and when it is, patients are often not properly informed, even when facing considerable levels of exposure. Examinations should be undertaken in a way that is transparent and accountable to the patient as well as to the professions. Valid consent requires that patients be given appropriate information. Disclosure should be full, frank, open, and include all material risks to which a reasonable person would be likely to attach significance. It must be presented in a way that the individual can assimilate, and be clarified by encouraging questions, which are answered honestly and completely (IAEA 2011, 2015; Malone et al. 2012; Bonn Call 2016).

Higher dose procedures (e.g. some CT examinations and interventional procedures, require open discussion and shared decision-making, something few departmental organisational arrangements are well equipped to deliver. The current situation in radiology is one in which communication is incomplete and/or unsuccessful. Simple matters of fact and probability are not known and/or not transmitted in an effective way to those who need to be aware of and have confidence in them. With widespread unawareness of radiological risk, the validity of consents, if obtained at all, must often be in doubt (Picano 2004a, 2004b). Failure to recognise and communicate potential risk can lead to social amplification of its perception. Ineffective communication and/or excessive reassurance may ultimately damage confidence in the professions and their processes (Bryman 2001; Picano 2004b; Malone et al. 2012).

It is never a matter solely for the doctor or professional to decide for another individual, except in circumstances where it is not feasible to obtain consent or where the risk is very small, and the consent is clearly implied by the circumstances. This is at variance with much of current practice, where the culture of radiation protection has come to rely on professionals avoiding talking to patients about the uncertainties involved, and assuring them that everything is fine. This is out of step with the emphasis now placed on the dignity and autonomy of the individual and the emphasis on honesty, which will be discussed later (Sia 2010).

Inevitably, these developments will place additional burdens on those involved in the practice of radiology, radiation oncology, and radiation protection. New operational approaches will be required to deliver an ultimately non-negotiable explicit or implicit valid consent. The imperative to provide information and/or obtain consent is not only an ethics one, it is also underpinned in several legal instruments and many judicial decisions. Many legal systems now tend to encourage and enable patients to make decisions for themselves about matters that intimately affect their own lives and bodies. The IAEA BSS requires that a procedure not be carried out unless 'the patient has been informed, as appropriate, of the potential benefit of the radiological procedure as well as radiation risks' (Picano 2004b; IAEA 2011, 2014; Malone et al. 2012; Malone 2014).

Much of the above comment is related to diagnostic imaging. However, radiotherapy is also home to much uncertainty, some of it related to problems with the quality of the evidence base for treatments. As a discipline, it draws heavily on technological and scientific advances, and radiation oncologists often find themselves offering treatments based on their own experience or a relatively slight evidence base (Chetty et al. 2015). When the evidence base is not conclusive, the dignity of the patient requires full disclosure of the uncertainties involved, both to allow them to make a good decision and to give real informed consent.

3.3.2 Non-Maleficence and Beneficence

Within medicine, there are longstanding formal commitments to both non-maleficence and beneficence. For example, not doing harm is one of the central features of the traditional Hippocratic Oath, and working for the health and well being of the patient is featured in the Declaration of Geneva. With respect to radiological protection of the patient, non-maleficence and beneficence together support both justification and optimisation. A rather simplistic view of non-maleficence would see it as closely aligned to optimisation, as manifested in the ALARA principle (Section 3.2.2.2 above). In practice, optimisation and ALARA are more complicated, as the interests of the wider community must also be accounted for, including economic and societal considerations. This renders the problem of balancing the various competing values and interests even more complex and will be further discussed in Chapter 6 under the heading 'Solidarity'.

In medicine, there are situations where pain or injury must be inflicted to achieve healing. This, in principle, is well accepted and not subject to serious debate. It is a matter of balancing the therapeutic outcome against the harm (i.e. non-maleficence and beneficence need to be balanced). Radiotherapy readily adopts the doctrine of double effect, allowing for the inevitable harm to the patient's healthy tissue to address life threatening conditions. This precedent, however, does not quite apply to diagnostic imaging, which is not *per se* therapeutic. Rather it is investigative, with the associated challenge of inflicting possible or probable serious harm.

The possibility of harm arising from diagnostic imaging provides a good example of situations where the requirements of dignity and autonomy need to be balanced against the requirement to do no harm. For example, take a situation where an individual strongly wishes to have a CT scan performed as part of a check-up, even though there are no risk factors or symptoms to suggest it will be of any value. At a simple level, dignity and autonomy would suggest that his/her wish be respected, and the scan be performed, particularly if public resources are not involved. On the other hand, the fact that probable harm will result from the scan suggests it should not be done. The other values also enter into this example, with prudence suggesting it would be better if it is not done, and justice suggesting it may be a bad use of a limited resource and so on.

A central issue for patients is that procedures should be conducted competently using optimised protocols/equipment. In countries with reasonable education and training for radiologists and technologists, the former tends not to be a major problem in practice. The latter, a real ethical concern, can be overlooked. A particular problem arises with the knowledge base out of which doctors and other practitioners are operating. This impacts on all five values, but we include it here for convenience. There is an extensive published literature in the area since 2004, which indicates that many (sometimes an overwhelming majority) of referring physicians and imaging practitioners have limited awareness of the actual radiation doses and risks involved in diagnostic imaging. Numerous publications from all corners of the world have established that few of those responsible for prescribing or performing examinations are familiar with the units used to specify the amount of radiation involved, the probable risks associated with it, and the relative risk rating of different examinations. While there are exceptions to this situation, and while there has been some improvement in some countries, the knowledge base out of which health professionals are operating continues to be unsatisfactory (Shiralkar et al. 2003; Lee et al. 2004; Krille et al. 2010; Singh et al. 2015; Campanella et al. 2017; Semghouli et al. 2017).

3.3.3 Justice

The third value in the pragmatic set, Justice, has several important consequences in radiological protection. As discussed in Chapter 2, it is the primary foundation of the concept of dose limitation. The most widely recognised forms of dose limitation are the legal dose limits prescribed for workers and the public. As noted earlier in this chapter, dose limits are not applied to exposures regarded as medical. However, two other forms of *de facto* dose limitation are applied (i.e. dose constraints and DRLs in the interest of non-maleficence as well as justice).

Dose constraints are values lower than the dose limits often used, for example, for planning a building, to ensure that the project will not only meet the dose limits but will also be optimised. DRLs are non-mandatory local or national dose values for particular radiological examinations against which aspects of local performance can be evaluated (ICRP 2007b). Greatly exceeding DRLs may be due to poor technique or protocols and be amenable to improvement through improved training and management. However, it may be more difficult to improve performance if the problems are attributable to equipment that needs to be updated or replaced (EC 2012). It is common to apply dose constraints and DRLs to the planning of radiological suites and the protocols for using them, but practise in this regard differs greatly throughout the world (EPA 2009). There is much scope for applying the value of justice to drive improvements in these areas.

The requirement for safe equipment and buildings arises under all five values in the pragmatic set and is discussed here due to its connection with dose limitation. It is recognised that ethical performance of examinations requires that they be performed on equipment that is acceptable from the point of view of patient safety and adequate performance of the examination. This includes the requirements that the equipment be capable of being optimised and of delivering examinations or treatments within the range of diagnostic reference levels that prevail in the region/country in which it is installed. In the EU, criteria of acceptability of radiological, nuclear medicine and radiotherapy installations have been established to assist ensuring equipment is performing satisfactorily, and detailed advice in the area is provided in a publication on the matter (EC 2012, 2015b). Regarding buildings, dose limits and/or dose constraints provide the basis for their design so that those working in them or visiting them as patients or non-radiation workers will not receive doses more than prescribed limits or constraints (EPA 2009).

Justice implies an equitable distribution of resources and thus does not favour excessive resource utilization on inappropriate, unnecessary examinations. Where the overall availability of resources is limited, as it is in diagnostic imaging and radiotherapy, justice has additional important consequences. Unnecessary and inappropriate deployment of the resource is not just wasteful, it also deprives persons in need access to the resource. In the case of radiology, such an example may be found in individual health assessment (IHA) of asymptomatic people, where examinations challenging traditional justification may displace examinations of patients in genuine need (Chapter 4). This can be further exacerbated by follow up of incidental findings in the IHA investigations, which displaces even more worthy cases. IHA is sometimes viewed as having the quality of an adverse event that deprives a person who really needs the examination of the opportunity to have it (Malone et al. 2016; Papanicolas et al. 2018).

In radiotherapy, failures of justice in this sense also occur. For example, use of an unnecessarily complex procedure for palliative pain limitation, when a simple procedure would be adequate, could deprive a patient with real clinical need of access to the complex procedure (Chapter 5). On a larger scale, the scarcity of radiotherapy resources is dramatically evident in developing countries with few, if any, radiotherapy facilities. Distances from centres vary dramatically, with patients even in the most developed countries traveling for many hours for their treatment. The IAEA initiated the AgaRT (Advisory Group on increasing access to Radiotherapy Technology in low and middle-income countries) to address the global accessibility injustice in radiotherapy and cancer care more broadly (Abdel-Wahab et al. 2017). Fairness and justice are also relevant in addressing patient prioritization and waiting times.

In diagnostic imaging or radiotherapy, a service provider may inadvertently, or otherwise, be diverted from his/her main focus (i.e. the wellbeing of the patient). Financial interest in maximising the use of a clinic's resources may interfere with risk–benefit evaluation. When a physician has such a financial interest, it must be disclosed to the patient (IAEA 2009). The economic cost and the loss of benefit to those who really need the services have been well articulated during the earlier Obama health reforms and more recently (Wennberg et al. 2008; IAEA 2011; Choose Wisely 2018; Papanicolas et al. 2018). Over and above financial gain, numerous other motivations contribute to poorly justified, inappropriate radiology including, for example, provision of medico legal cover for the physician. The latter in its pure form is for protection of the physician and not for the benefit of the patient and hence does not constitute a *bona fide* justified medical exposure, even though it is not uncommon.

In all of the above, justice and individual autonomy must be balanced sensitively against each other. Finally, justice also implies a concern for fairness in our treatment of the most vulnerable, such as children, or radiation-sensitive individuals.

3.3.4 Prudence

One of the most discussed additions to Beauchamp and Childress values in medicine is Prudence (Beauchamp and Childress 2013; Zölzer 2013). As discussed in Chapter 2, it, or the related precautionary principle, has been embraced in several areas where scientific uncertainty impinges on public policy. It may be paraphrased by stating that where an action potentially causes a serious irreversible harm, measures to protect against it must be taken even if the causal relationships involved are not fully established scientifically (Malone 2013; Zölzer 2013; ICRP 2018). Prudence is an extension of beneficence and non-maleficence, and helps deal with the unsatisfactory state of our knowledge about radiation risks. Beauchamp and Childress do not widely invoke it as a value, but given the prominence of uncertainty in radiological protection, this value is potentially helpful. This is confirmed by its inclusion in both ICRP 138 and in the pragmatic set (Malone and Zölzer 2016; Zölzer 2013, 2017; ICRP 2018).

There is much confusion about the place of prudence in dealing with the risks from small doses of radiation in medicine, particularly in radiation protection of patients. However, for this purpose, clear, high-level precedents are available from the Wingspread Conference and the Rio Declaration following a major UN Environmental Conference (UN 1992; Wingspread Conference 1998). This is also sanctioned by the great philosophical and religious traditions as detailed in Chapter 2. Thus, prudence and the precautionary principle imply looking at the potential for serious harm before it happens, even where the evidence base is incomplete, and has been adopted at the highest possible levels globally. This has a valid and under-rated resonance in radiation protection of patients and workers.

On the surface at least, ICRP appears to support the precautionary principle, particularly in adopting the LNT model for extrapolation to low doses. Yet it also states '*... calculation of the number of cancer deaths based on collective effective doses from trivial individual doses should be avoided*'. This is justified by saying that such calculations would be '*biologically and statistically very uncertain*'. As the precautionary principles applies precisely to those cases involving uncertainty, the ICRP position seems somewhat self-contradictory. The United Nations Scientific Committee on the Effects of Atomic Radiation (UNSCEAR) position, discouraging population risk calculations for small doses, also requires more robust justification with respect to prudence and the precautionary principle as do restrictive practices on use of terminology in some scientific journals (UNSCEAR 2012; MP 2013; Zölzer 2013; Malone and Zölzer 2016).

The attitude of medical radiation protection practitioners to the precautionary principle has not been surveyed and is not known. However, the experience of the authors is that it varies greatly, ranging from ignoring the possibility of risk at one end to stating the probable risk as real and certain at the other. Similar ambivalence exists among the allied professions. In view of this, prudence/the precautionary principle need to be repositioned, and be applied more consistently in the practice of medical radiation protection (Malone and Zölzer 2016).

3.3.5 Honesty

Honesty, in the sense used here, is a procedural value, and extends well beyond financial matters. It is often presented as 'working in an open and transparent manner', and requires that people are not deceived. To paraphrase Chapters 2 and 6, honesty has been suggested as a guiding value for interaction between experts with specialist knowledge and lay people, a situation that precisely describes the interactions between patients and the health professionals involved in radiation protection in medicine. Thus, it is not surprising that honesty is included as the fifth and final value in the pragmatic set. As mentioned earlier, many of the major medical scandals of recent decades have been rooted, at least in part, in failures of honesty. This often involves lack of candour with information and an unwillingness to acknowledge the importance of extending accountability to patients, and not just to the professional peer group.

The history of the nuclear industry, and by association of radiation protection in general, is not one in which honesty and transparency could be regarded as stand-out features. The culture of radiation protection in medicine is sometimes prone to a sense of siege, and compensates by underplaying the probable risks and the need to communicate about them. This can seem like old-fashioned paternalism re-emerging in another guise. The obligation to communicate openly extends to areas where uncertainties prevail – in fact become even more important when uncertainties are involved. Closing off the uncertainties with an excess of reassurance, even in the face of explicit requests for information, is not a good solution. Thus, in terms of honesty, much remains to be done in radiation protection in medicine.

3.4 CONCLUSIONS

The pragmatic set of five values will help guide evaluation of day-to-day medical uses of radiation. The need for such a practical set arises from the high levels of expectation of behaviour of health professionals, particularly in the event of accidents or misadventure that are subject to public scrutiny. Three of the values – dignity/autonomy, non-maleficence/beneficence and justice – derive from the well-regarded Beauchamp and Childress approach, and are independent of ethical theories and acceptable in most cultures (Beauchamp and Childress 2013). The other two – prudence and honesty – are derivable from the Beauchamp and Childress approach, but are explicitly included to address practical concerns in areas where the day-to-day culture of medical radiation protection may be somewhat at variance with contemporary public expectations. This is particularly so when dealing with the uncertainties around radiation risks. All the values are included by ICRP in its assessment of the ethical basis of the system of radiological protection as a whole, though a difference in emphasis may be necessary to take account of the medical context. This is partly due the fact that the medical area still awaits the ICRP's full attention regarding its ethics basis (Malone and Zölzer 2016; ICRP 2018).

The importance of the two additional values – prudence and honesty – is often overlooked. At this stage they buttress the core Beauchamp and Childress values. When a fully developed and widely agreed system is well embedded in radiation protection in medicine, the need to state these explicitly may decline. But, in the meantime, prudence and honesty offer clear and reasonable guidance on how to behave, particularly in the absence of full knowledge.

The values of dignity/autonomy and honesty imply that patients have the right to know of possible risk, so that they can make good informed decisions about their own healthcare. Radiologists, radiation oncologists and other healthcare professionals have a duty to inform patients of benefits and risks, based on the best available estimates and the associated uncertainties. Routine excessive reassurance is not appropriate and ultimately damages credibility. Furthermore, patients constantly encounter and cope with larger uncertainties during their medical interventions. For example, some cardiac interventions carry risks of fatality or serious morbidity in the range 1%–5% for some interventions.

There may be incompatibilities between the systems of medical ethics and radiation protection. The alignment of ethical values underlying the practice of medicine and the ICRP's core principles: justification, optimisation and dose limitation has not been fully explored, is a problem for both, and a task for the future. However, because of their basis in medical ethics and social expectation, the pragmatic value set can be expected to provide a good interim approach and achieve a higher level of recognition in medicine than the ICRP principles on their own. The approach might also facilitate making the core messages of radiation protection better known to its largest area of application.

In conclusion, the factors determining outcomes and ethical behaviour are, as we have seen, many faceted. There are many areas of radiology, dentistry and radiotherapy in need of attention from an ethical point of view.

The values should help movement toward a style of behaviour in radiation protection that is consistent with good medical practice and with contemporary social and ethical thought. Clinical disciplines will be enriched by a more explicit commitment to practice informed by the pragmatic set. As will be seen, it has proved to be an effective roadmap in evaluating the twenty or so scenarios in Chapters 4 and 5, as well as elsewhere throughout this book.

CHAPTER 4

Ethics Analysis of Imaging Scenarios

4.1 INTRODUCTION AND METHODOLOGY

Chapters 1 and 3 made it clear that the prevailing social environment has raised public expectations for medical imaging, including radiation protection. Not least, in this respect, is the level of transparency, accountability, and prudence expected of professionals and institutions. Chapter 3 summarised the legal, professional and ethical aspects of radiation protection in medicine that would render it compliant with the ICRP framework, medical ethics and social expectations.

Radiation dosage with plain radiography was, for many years, regarded as a non-issue. Whether or not this was the case, is debatable. However, during the last 10–15 years, doses have increased to worrying levels, to an extent that their containment has become a competitive marketing tool. The largest component of the dose involved arises from computed tomography (CT) which is now widely deployed. Each examination should, in theory, provide a diagnostic benefit, whether performed in high or low-income countries, the public sector or the private domain.

Each examination represents a monetary and opportunity cost/risk, which the patients and doctors may, or may not, be aware of. While dose will continue to be an issue, recent developments in CT promise reductions that remain to be fully evaluated. It is to be expected that performance of examinations and the processes/protocols surrounding them will be respectful of the pragmatic set of values outlined in Chapters 2 and 3 (Table 3.1) above, for example, they will:

- Respect individual dignity/autonomy
- Do no harm and some good
- Be consistent with justice
- Be performed with prudence and precaution
- Be consistent with honesty and transparency

Ten scenarios follow, drawn from diagnostic imaging practice, and the extent of compliance with these values is assessed in each. This should assist the reader in visualising how the pragmatic value set can be applied in practice. It will also give some insight into the extent to which the pragmatic set overlaps with the principles of the ICRP when judging specific situations.

The scenarios involve individual examinations or processes, generally conducted in a medical setting. The intention is to illustrate the application of the values in plausible diverse situations, indicate how they might be deployed, and stimulate further much-needed work in the area. This inevitably involves some over-simplification and developing scenarios in which negative features arise. However, it is not our intention to be unduly critical or to offer prescriptive remedies. Effective solutions will inevitably have to be evaluated and assessed in the real world of departments delivering day-to-day service. Instant solutions proposed here could well prove facile.

Each scenario is described and then scored as complying (or not) with the values in the pragmatic set. The evaluation of compliance, or otherwise, is the personal judgment of the authors. Compliance with a value is indicated as being strong **(Y)**, weak **(y)** or neutral **(–)**. Likewise, Non-compliance is indicated as strong **(N)**, weak **(n)** or neutral **(–)**. Some aspects of scenarios demonstrate compliance with a value, when considered from one perspective, and non-compliance, when considered from another. Thus, it is possible to score both **(Y/y)** and **(N/n)** for the same value. While, this may be troubling to some, it should not be difficult for physicists whose discipline is home to the uncertainty principle and the elegant if confusing states of matter central to Quantum Mechanics. Earlier versions of Scenarios 1, 3, 6, 8, and 9 were originally published in the *British Journal of Radiology* (Malone and Zölzer 2016).

4.2 SCENARIOS

4.2.1 Scenario 1: Professor Browne, Orthopaedic Surgeon

Prof. Browne is a professor of orthopaedic surgery. He holds a weekly outpatient clinic in a public hospital, where he sees both new patients and follow-up cases. His clinic is well resourced and is a model of efficiency, running to time with little waiting around for patients. Prof. Browne insists all patients attending have an up-to-date radiology examination of the relevant part before he sees them. This obviously challenges the principal of justification. The director of radiology and the MPE (medical physics expert, see Chapter 3) advised him against this practice. His response is dismissive, pointing out that radiology in their hospital is home to queues and waiting lists, whereas he runs an efficient patient-friendly service.

He requires that patients bring their film folder or DVD to the clinic. He states it takes too long to get a radiology report which, when received, may not address the issue he is interested in. So, he reads the images himself. Pre-signed forms or authorised referrals on the information system are provided for patients, and they are sent to radiology for the required examination. The Radiology Department is concerned that Prof. Browne may bypass the department entirely if he is refused, so it reluctantly accepts the situation. Prof. Browne doesn't share any of these concerns with his patients. Likewise, he doesn't discuss benefit/risk information with them, which he dismisses as largely speculative. He feels that as a doctor, his role includes being an advocate for his patients, and acts accordingly.

Prof. Browne is obviously of the paternalist school. The two-row table at the bottom of Table 4.1 indicates how well his practice complies

TABLE 4.1 Scenario 1: Professor Browne, Orthopaedic Surgeon

- Orthopaedic surgeon Prof Browne, weekly public clinic in large hospital.
- New, follow up, injuries and elective patients.
- Insists all patients sent to radiology by the nurse.
- Will not see a patient without a film folder or DVD from radiology.
- Refuses radiology/physics justification advice.
- Proud of efficient patient-centred service.

Dignity Autonomy	Beneficence, Non-Maleficence	Justice	Prudence Precaution	Honesty Transparency
(y)	(–)	(y)	(–)	(–)
(N)	(N)	(N)	(N)	(N)

with the five-value pragmatic set. Clearly, the practice fails all five on significant grounds, and he scores a no **(N)** for each. With respect to dignity/autonomy, this is not respected in the way the decision to conduct examinations is taken, so this scores **(N)**. He also scores a small yes **(y)** in recognition of his efforts to provide a timely, efficient service respecting his patients. His practice exposes many patients to unnecessary radiation risk with no benefit, so he scores an **(N)** under non-maleficence. He reads the images himself which, some will argue, adds to the potential for harm. He scores **(N)** under justice as either the patient, insurance or society will have to pay for all unnecessary and possibly useless examinations. His timely, efficient service also scores **(y)** under justice. He does not reckon on the possibility of risk and offers practically no information to the patient, so scores a clear **(N)** for the last two headings: prudence and honesty.

4.2.2 Scenario 2: Mr Grey, Ultrasound Referral

This scenario deals with a case in which Mr Grey is referred by his general practitioners (GP) for an ultrasound examination to explore the cause of ongoing upper abdominal pain. The GP has reason to suspect gallstones as the cause but omits to include this in his referral note. Mr Grey is chairman of the hospital board and staff in the imaging department greet him on arrival. They decide to add an abdomino-pelvic multiphase contrast CT examination to the ultrasound scan to provide their chairman with the most discerning service they can offer. The staff felt that would move quickly to a diagnosis and staging, if something like cancer were involved. They might not have taken this approach had the referral note been more complete and mentioned the GP's suspicions. The radiation and other risks of this high-dose CT procedure are explained to Mr Grey and he consents to it.

The dose noted for Mr Grey is in the upper end of the range for CT examinations. This is not warranted, as the simpler ultrasound procedure, performed first, could have identified gallstones as the source of the problem. Both examinations were performed, and the CT confirmed the ultrasound diagnosis. Within the ICRP system, the problems here include failures of both justification and optimisation. They arise from several sources including, in the first instance, an inadequate referral note from the GP. Had he mentioned his suspicion about gallstones, the more elaborate CT scan might not have been undertaken. There could also be inadequate protocols for identifying the correct examination in radiology. However, in this case, the protocols were satisfactory and would have directed Mr Grey first to ultrasound. But, the staff were anxious to offer

TABLE 4.2 Scenario 2: Mr Grey, Ultrasound Referral

- Mr Grey referred for ultrasound for upper abdominal pain.
- GP suspects gallstones but does not mention this in referral.
- Mr Grey is Chairman of hospital. Staff add complex CT Scan. Risk from CT is explained, and consent is obtained.
- Complex CT inappropriate according to guidelines.
- Excellent scans performed promptly, revealing gallstones on US followed by CT.
- Staff pleased they gave their chairman of their best.

Dignity Autonomy	Beneficence, Non-Maleficence	Justice	Prudence Precaution	Honesty Transparency
(y)	(–)	(–)	(–)	(y)
(N)	(N)	(N)	(N)	(n)

their chairman the most sophisticated examination that might help, and thus included the elaborate examination, which proved quite unnecessary.

In getting Mr Grey's consent, the explanation offered to him referred primarily to dose and risk. It did not refer to the fact that the examination would not be appropriate, based on most guidelines in the area, and the much simpler ultrasound would have been adequate initially (RCR 2018).

In Table 4.2, respect for the dignity/autonomy of the individual is recognised through explaining the risk and obtaining consent, but only receives a small **(y)** as the inappropriateness of the examination was not mentioned. The more complex CT might only have been necessary if cancer had been established and was being staged **(N)**. The consequences were inadequately recognised by the staff in their anxiety to do well for their chairman and gave rise to the problems noted. These include failures under the headings of justice **(N)**, as the complex CT is a poor use of resources. In addition, there are clear **(N)s** under non-maleficence and prudence because of unnecessary exposures and the associated probable risk of harm. A **(y)** under honesty is scored for presenting accurate information on risk etc., and an **(n)** for inadequately presenting the information about appropriateness of the CT examination to the chairman.

4.2.3 Scenario 3: Dr Pine, Paediatric Radiologist

This scenario concerns Dr Pine, an experienced, well-trained paediatric radiologist. A two-year-old boy is referred for a whole-body CT examination. Dr Pine believes the examination is justified. She advises the child's parents, his legal proxies, that the examination should proceed. The parents enquire about the risks, if any, from the examination. Dr Pine reassures them that there are none they need to consider. She deflects further questioning by

TABLE 4.3 Scenario 3: Dr Pine, Paediatric Radiologist

- Two-year-old boy presents for whole body CT examination.
- Dr Pine, well-qualified paediatric radiologist, assesses situation and believes examination is justified and should be performed immediately.
- Parents request information on cancer risks. Dr Pine deflects questions, explaining her department is the best in the country for this type of case.
- Parents unhappy with response, but consent.
- Her reasons for doing so are that full explanation takes too much time, and fear the parents may withdraw the child from a necessary examination.
- A technically excellent examination is performed.

Dignity Autonomy	Beneficence, Non-Maleficence	Justice	Prudence Precaution	Honesty Transparency
(Y)	(Y)	(Y)	(Y)	(–)
(N)	(–)	(–)	(–)	(N)

explaining that the hospital is the leading one in the country for this type of examination in young children (which is true) and it will not be better performed elsewhere. Her reasons for deflecting the question, which she does as a matter of policy, are two-fold. First, it takes too much time to respond to detailed requests for further information. Second, and more important in her mind, informed parents may withdraw their child from what she believes is a necessary examination. The examination is clearly justified from the history provided by the referrer and is technically well performed and reported on promptly.

This scenario raises interesting problems illustrated in Table 4.3. Clearly the dignity of the child is respected in ensuring the examination is justified and his legal proxies have given consent **(Y)**. However, the radiologist's behaviour towards the parents does not respect their dignity/autonomy and role as legal proxies for the patient by failing to provide the information necessary to allow them give a truly informed consent **(N)**. Behaviour with respect to the honesty category was also unsatisfactory. Patients or their legal proxies are entitled to, and should receive, honest transparent information, when they request it **(N)**. The other categories, non-maleficence, prudence and justice were all exemplary and, hence each scores a yes **(Y)**.

Medical exposure of children generally requires special considerations as they have higher radiation sensitivity than adults and a longer life expectancy, which gives added emphasis to the values of non-maleficence and the principle of justification. Thus, imaging techniques that do not use ionising radiation should be considered, where it is possible to substitute them without loss of diagnostic information. Consent must be obtained, and where a child cannot give it, it should be secured from a lawful proxy. Facilities undertaking radiology of children should have paediatric protocols as adult protocols will often give excessive dose to children. It is important to note the need for special justification and optimisation because of the different patterns of presentation of paediatric diseases, and the obvious smaller sizes. Additional special considerations arise where the possibility of abuse of the child and/or his/her siblings arises.

4.2.4 Scenario 4: Mr Viridian, Mistake in Referral

Mr Viridian attends the Nuclear Medicine department as part of the follow up of his prostate cancer. His urologist, Professor Coral, feels a bone scan is needed as part of the investigation. He asks a junior doctor, who is part of his team, to attend to the request and to check the indications for a bone scan provided by the Nuclear Medicine Service. The junior doctor is under pressure and later checks the indications and finds the scan is appropriate for Mr Viridian. He completes the request procedure, giving a short clinical history. Mr Viridian attends the Nuclear Medicine department some days later and a scan is performed in a technically excellent manner.

The report on the scan is returned to Prof Coral reassuring him that the patient's lungs are normal and there is no sign of a pulmonary embolism. He consults the patient's record and finds, to his surprise, that a lung scan had been requested by the junior doctor. On consulting the request form, he also finds that the clinical details given were consistent with the need for a bone, not a lung scan.

After investigation, it is found that the junior doctor had two nuclear medicine scans to request, one for a lung scan and one for a bone scan, each for a different patient. He was distracted by a phone call as he completed the requests, resulting in Mr Viridian having a lung scan requested instead of a bone one. The other patient, vice versa, had an inappropriate bone scan. In both cases the clinical details given on the forms were appropriate to the correct scan and would have raised concerns had they been reviewed in the Nuclear Medicine department. Following the investigation, Prof Coral and Dr Burleywood, the physician in charge of nuclear medicine, decide that the incident is an unfortunate accident involving both their services. They decide not to tell Mr Viridian, and the other patient – who had been incorrectly injected – as it might upset them. Instead they request an additional scan for each patient without charge. They also decide to institute measures to improve the attention to clinical details on the request forms for nuclear medicine scans.

TABLE 4.4 Scenario 4: Mr Viridian, Mistake in Referral

- Mr Viridian attends nuclear medicine for a bone scan, as part of follow up of his GU cancer, organised by urologist, Dr Coral.
- A lung scan was incorrectly requested and performed.
- However, clinical details provided appropriate to a bone scan.
- Another patient involved, incorrectly received the bone scan.
- Dr Coral and the head of nuclear medicine, Dr Burleywood, decide not to tell the patient, and not to report to authorities.
- Repeat scans are performed without charge to the patients.

Dignity Autonomy	Beneficence, Non-Maleficence	Justice	Prudence Precaution	Honesty Transparency
(y)	(y)	(y)	(–)	(–)
(N)	(N)	(n)	(N)	(N)

In Table 4.4 we have a situation where potential harm to both patients was incurred and scores a **(N)** for non-maleficence, but also a **(y)** for the attempt to correct the error and do what the situation demanded immediately. Likewise, both **(y)** and **(N)** are scored under dignity and autonomy in respect of the care accorded in ensuring the bone scan was justified, and the failure to respect the patient's right to know they had been the subject of an accident. Similarly, a **(N)** is scored under honesty for failure to be candid with the patients, and for failures to report as appropriate to regulatory bodies or the hospital authorities. The precautionary principle is violated in the nuclear medicine department in not having a system that better checks the clinical information provided against the requested scan in both instances. Not charging the patients for the repeat scans scores a **(y)** under justice, whereas the fact that the repeats are performed at public expense incurs an **(n)**.

In this case, there are good practice and legal issues that will vary from country to country. The initial fault is the incorrect referral, compounded by the failure to note the lack of correspondence between the clinical notes and the scan requested. Both are medical errors and may require reporting to the medical authorities. If, on the other hand, the correct scan had been requested with the correct clinical details, and the error lay in the wrong injection being given leading to the wrong scan being performed, this might be a radiation regulatory matter. Perhaps surprisingly, much of the ethical analysis above would remain unchanged as much of the problem lies in the incorrect injection, and the attempts to protect the patient from information about what happened. If the scan dose (1–2 mSv) exceeded the annual dose limit for a member of the public (1 mSv), as it could in this case, an excessive exposure of a member of the public may have to be reported to the radiation regulatory authorities (ICRP 2017; RCR 2017; SNMMI 2018). There are many permutations and combinations of the types of error that can occur and the different obligations they place on those involved.

4.2.5 Scenario 5: CT Dose Dilemma

St Elsewhere's Hospital is a public facility with an adjoining private hospital where doctors from the public facility can also practice. The public hospital has older CT facilities with multislice functionality but does not have sufficient capacity to meet its patients' needs. Excess demand is referred to the private hospital which is equipped with the most up-to-date CT and MRI equipment, the former having low-dose facilities and dose sparing protocols for adult and paediatric patients. The older public facilities can give doses in the range 2–10 times those in the private, for similar clinical examinations.

The radiology department is committed, in both the public and private facilities, to the most up-to-date good practice, including selective self-audit of differing aspects of the departments' work each year. The most recent audit indicates a strong selection of older patients being referred to the private lower-dose facility for CT scans. Further investigation of this phenomenon reveals that older patients also tended to have better, more comprehensive private insurance than younger patients. It also reveals management pressure in the private facility to encourage referral of patients with good private insurance in preference to the less lucrative public referrals.

Table 4.5 reveals that none of the values in the pragmatic set favour the approach identified in the audit and the follow up investigations, even

TABLE 4.5 Scenario 5: CT Dose Dilemma

- St Elsewhere's, a public facility, adjoins a private hospital.
- Both have CT scanners. The equipment in the private is newer and has better low-dose facilities.
- Public hospital lacks capacity for its imaging needs, and some patients referred to the private hospital for CT imaging.
- Audit shows older patients preferentially referred private.
- Further investigation indicates older patients also have better private insurance.

Dignity Autonomy	Beneficence, Non-Maleficence	Justice	Prudence Precaution	Honesty Transparency
(–)	(–)	(–)	(–)	(–)
(–)	(N)	(N)	(n)	(N)

though they provide a technically good and much more timely service to the patients than would be otherwise available. With respect to individual dignity/autonomy, the preferential patient referral only became evident when averaged financial data were looked at. Individually all the referrals appear, on the surface, to be reasonable. Should this practice, once identified, continue, it would score a **(N)** under dignity and autonomy. However, with respect to the other four pragmatic values, non-compliance **(N)** is scored in all cases, and there are no counterbalancing positive scores. Clearly there is probable harm with the preferential receipt of higher doses by the younger population; equally clearly this is unjust in terms of equity between age groups; and there is a lack of openness and transparency in that the patients and many professionals in the hospitals are not aware of the situation. Finally, there is some lack of prudence/precaution involved in allowing this situation to come about and persist.

4.3 SCENARIOS: SPECIAL CONSIDERATIONS FOR PREGNANT PATIENTS

Medical exposure of pregnant or potentially pregnant patients gives rise to additional ethical considerations. In evaluation of risks and benefits from medical exposures during pregnancy, both the mother and the embryo/foetus must be considered (Wagner et al. 1997; Malone 1998; EC 1998a; ICRP 2000a; Faulkner et al. 2001). The mother may receive direct benefit while the foetus may be exposed without direct benefit arising from its own exposure (or vice-versa). On the other hand, if the mother's problem is life threatening, the procedures involving her radiation exposure may contribute to her survival, which in turn would directly benefit the foetus. In this setting, the mother has a role-related responsibility to care for her unborn child as well as to make decisions for herself. The pregnant patient has a right to know the magnitude and type of potential radiation effects that might result from in-utero exposure, and her fully informed consent assumes additional important features because of this. Many countries, and individual hospitals, have policies to avoid irradiation involving the foetus arising from elective diagnostic imaging (Schreiner-Karoussou 2009; ACR 2018). Where it proves necessary to use radiation during pregnancy, a risk assessment should generally be conducted. Each case must be reviewed according to the gestational age at the time of exposure, the extent to which the foetus is included in the radiation field, and the magnitude/distribution of the radiation doses received.

The situation is much more difficult when the patient is not sure or does not know that she might be pregnant. In such situations, the EC requires that where pregnancy cannot be excluded it should be assumed (EC 1998a; EC 2013). The degree of concern and form(s) of consent that should be obtained vary greatly from country to country and are dependent upon the extent of the risk to the foetus and mother. In such circumstances a risk assessment should be undertaken, and the patient should be counseled accordingly by a knowledgeable, experienced person that she can trust. Where inadvertent irradiation of a pregnant patient takes place, a risk assessment, including the risk to the foetus, should also be undertaken, and again the patient must be sensitively counselled. Two scenarios, 6 and 7, dealing with these issues follow. Pregnancy is also incidentally part of Scenario 9, later in this chapter.

4.3.1 Scenario 6: Ms Violet, Pregnant Patient

Ms Violet, a female patient in her mid-30s, is referred to the radiology department of a moderate-sized district hospital for an elective non-urgent pelvic CT scan. Her medical history based on the family physician's referral note justifies the scan in the opinion of the radiologist. An appointment is arranged some months ahead, as she will be away on holiday in the meantime. She attends the hospital at the appointed time, and is asked if she is pregnant, or might be pregnant, to which she responds 'no' as she 'is careful'.

Ms Violet has the examination, which reveals no pathology, but also discloses that she is pregnant. She is delighted to find herself unexpectedly pregnant and can't explain how she didn't notice. But she is also worried and confused about the situation and by the possibility that the child she is carrying may have been damaged by the scan. She reviews the information available on the Internet and finds the US Food and Drug Administration attributes a potential risk of future cancer to irradiation. She finds the level of risk suggested unacceptable. Later Dr Matte, the radiologist dealing with the examination, meets with her, assures her there is no significant risk for her to worry about, and advises that she should not be concerned. After further researching the

TABLE 4.6 Scenario 6: Ms Violet, Pregnant Patient

- Department policy: Ask patient at reception if she is pregnant.
- Does not have more rigorous approach as this would be 'time consuming' and/or 'inconvenient'.
- Ms Violet has elective pelvic CT and is found to be pregnant.
- She is happy to be pregnant, but very distressed about the radiation, does not trust the advice she receives, and considers seeking a termination.

Dignity Autonomy	Beneficence, Non-Maleficence	Justice	Prudence Precaution	Honesty Transparency
(y)	(y)	(–)	(–)	(y)
(N)	(N)	(–)	(N)	(N)

issue on the Internet, the patient finds what the radiologist stated to her unconvincing. She fails to attend a further appointment and decides to seek a termination.

Table 4.6 scores the extent of compliance with the pragmatic value set. On the positive side is the fact that the hospital asks the pregnancy question, and thereby tries not to do harm. In doing this, it aligns with much of conventional practice. On the negative side, as almost any member of the public could point out, asking someone if they are pregnant in a relatively public place is an unreliable method of establishing their status. Nevertheless, the hospital feels it is following available advice for good practice. However, while this advice claims it is based on current scientific evidence, it does not clearly acknowledge its dependence on value judgments. Hence, some pregnant patients and some members of the public could find the positions taken to be, for example, lacking in both transparency and prudence. The hospital's reasons for not offering a more rigorous assessment of potential pregnancy are that it is time consuming and inconvenient. The hospital has been challenged on the practice by previous patients, and after review, felt it would be too disruptive to alter it. This approach can be faulted on the grounds of failing to respect the dignity and/or autonomy of both the mother and possibly the embryo/foetus **(N)**; exposing both to potential harm **(N)**; failing to act prudently and follow the precautionary principle when there is possible but unproven risk **(N)**; and not behaving in a transparent way both before and after the examination **(N)**. Under the justice heading, the behaviour of the hospital might be taken as relatively neutral. A **(y)** is scored under dignity/autonomy, non-maleficence and honesty, for asking the pregnancy question, being willing to act on it and being open/transparent within the limits of the professional advice available to it.

The hospital's approach is consistent with much of the practice throughout the world. However, many of the areas in which medicine has found itself involved in public scandals are those where individual dignity and autonomy has been sacrificed to long-established and professionally sanctioned practices (Malone 2008, 2009).

4.3.2 Scenario 7: Ms Magenta, Pregnant Patient

Ms Magenta, aged 40, phones her local hospital for an appointment for an elective abdominal CT scan following referral by a gastroenterologist at the hospital. She is anxious about the result and explains she would like to have the scan soon prior to going on holiday and is given an early appointment. When she attends the hospital, she is asked, at reception, if she is pregnant and replies 'No'. On questioning, she states her periods are irregular, and have been so for many years. The hospital decides to proceed with the examination based on her sense of urgency, her history, and her confidence that she is not pregnant.

Ms Magenta has also had difficulty with conception, is having ongoing IVF treatment, and had fertilised eggs implanted around, or before, the time of the CT scan. She did not mention this at reception, as she wasn't asked, is sensitive about it, and furthermore, she assumed on the basis of prior experience that the attempt had been unsuccessful. Shortly after the CT scan, she visits her obstetrician to confirm this, but the obstetrician feels it may well have taken and she is probably pregnant. She is delighted with this news until a friend explains that if pregnant the CT scan could have damaged the embryo. She is distraught at this prospect, and arranges an appointment at the hospital to discuss the matter, explaining that she might have been pregnant at the time of the scan, but omits mentioning she was on IVF. She meets Dr Beaver, the radiologist, who advises she should not be concerned. Dr Beaver further explains that even if the embryo had been damaged, it would not have implanted and would have been lost, thereby ensuring there would be no prospect of a damaged child. Ms Magenta is so distressed by this that she leaves the interview before it is finished. Some days afterwards she has a heavy period confirming loss of the embryo. She consults the Internet about the matter and is further shocked to find the radiologist's advice echoed on several websites (Wagner et al. 1997; Malone 1998; HPA 2009).

TABLE 4.7 Scenario 7: Ms Magenta, Pregnant Patient

- Ms Magenta, aged 40, attends her local hospital for an elective abdominal CT scan.
- She is asked if she is pregnant and replies No. On questioning, she states her periods are highly irregular. The hospital decides to proceed with the examination.
- Ms Magenta is having ongoing IVF treatment, but does not reveal this, as she is not asked.
- A friend explains that if pregnant the scan could be damaging.
- Advice she receives from hospital, and various websites, shock her.

Dignity Autonomy	Beneficence, Non-Maleficence	Justice	Prudence Precaution	Honesty Transparency
(y)	(–)	(–)	(–)	(–)
(N)	(N)	(–)	(N)	(N)

Table 4.7 awards a **(N)** for respect for dignity and autonomy, mainly based on the inadequacy of the radiologist's interview after the event. However, the hospital systems score a **(y)** for asking the pregnancy question. Given how common infertility treatment is in women in Ms Magenta's age group, it might be expected that enquiries in that regard would be made, and the radiation protection policy would be adjusted to it. The possibility of harm was neglected, so a **(N)** is scored under both non-maleficence and prudence/precaution. However, within the culture of the radiation professions, the problem involved is not widely recognised or accepted. Finally, under openness and transparency, the hospital and Dr Beaver score a **(N)**, although Ms Magenta's lack of full disclosure is also to be faulted.

Issues around exposure of patients who are pregnant (as well as pregnant staff) provide examples of scenarios that can be better analysed and resolved in the context of the wider deliberation offered by the pragmatic value set, as opposed to that based on the legal/ICRP system alone. This is further illustrated in Scenario 9 below. Both approaches afford interesting problems for reflection/analysis, but the broader perspective of the pragmatic set helps the problem be viewed more holistically. This is particularly so in the case of irradiation during early pregnancy, as illustrated in Scenarios 6 and 7.

4.4 SCENARIOS: SPECIAL SITUATIONS: IHA, NON-MEDICAL EXPOSURES AND EQUIPMENT FAILURE

4.4.1 Scenario 8: Dr Salmon, Cardiologist in Private Practice

Throughout the world, a practice exists of screening asymptomatic individuals, without a relevant risk profile, with the most sophisticated of radiological examinations. This is referred to as Individual Health

Assessment (IHA) and is distinct from formally approved screening programmes, such as those for mammography and some other cancers (Malone et al. 2016). Frequently there is no evidence that IHA can be of value to those who avail of it, usually in the private sector. Examples include some instances of coronary artery calcium scoring, and some screening for cancers of the lung, colon, or abdominal cavity in populations without a predisposed risk profile. Some IHA challenges the ICRP principle of justification and the ethics values of non-maleficence, prudence, justice and honesty. Concerns about IHA arise from its unfavourable balance in radiation benefit/risk trade off. Both European and international regulations/standards address justification of radiological examination of asymptomatic persons (EC 2013; IAEA 2014; Malone et al. 2016). Fear of future disease is often a powerful influence. Advertising can play on fear, exaggerating benefits/convenience, and failing to mention associated risks.

Table 4.8 deals with the practice of Dr Salmon, an interventional cardiologist in private practice. She undertakes IHA for symptom-free patients with or without known risk factors, referred by other practitioners, self-referred, or who self-present (Chapter 3). Dr Salmon explains all the risks of interventional cardiac procedures including the potential radiation risks. She also qualifies this by explaining the radiation risk is unproven. She conducts the procedures on request and with formal consent. Separate fees are charged for the consultation and for the procedure. The procedures are undertaken in the associated imaging centre, in which she is a shareholder. Her financial interest is not disclosed to the patient.

TABLE 4.8 Scenario 8: Dr Salmon, Cardiologist in Private Practice

- Dr Salmon, Interventional Cardiologist. Private rooms with associated imaging facility.
- Explains the radiation (and other) hazards of procedures. Explains radiation risk is unproven.
- Accepts IHA and unreferred worried well.
- Procedure on request with consent.
- Fee for consultation, separate charge for imaging.
- Dr Salmon is shareholder in imaging facility and does not advise patients of her financial interest.

Dignity Autonomy	Beneficence, Non-Maleficence	Justice	Prudence Precaution	Honesty Transparency
(Y)	(-)	(-)	(–)	(Y)
(n)	(N)	(N)	(N)	(n)

In terms of compliance with the pragmatic value set, Dr Salmon scores highly on respecting the autonomy of the individual **(Y)** and on honesty **(Y)** as she takes a lot of trouble to inform the patient and get consent. She also scores an **(n)** for dignity/autonomy as she does not share the uncertainty about risk with the patient and an **(n)** for honesty, arising from non-disclosure of her shareholder interest in the imaging facility. On the other areas, including non-maleficence and prudence, she scores **(N)** arising from the probability that harm may flow from the unjustified examinations. She also scores a hard **(N)** under justice, as the examination is a poor use of a limited resource, and possibly excludes somebody who really needs it.

An important consideration in IHA is the cost of examinations in general and how they can be best deployed in the healthcare system. IHA may lead to false positives and incidental/equivocal findings. The cost and consequences of dealing with these can be significant and often fall on the public health sector, where such exists. If IHA is to be part of the healthcare system, it will require some adjustment to how it functions now. While Dr Salmon's private cardiology practice has many commendable aspects, it lacks an ethical and governance framework that can ensure it operates in the best interests of those who avail of it and the communities that host it (Malone et al. 2016).

4.4.2 Scenario 9: Ms Auburn, Drug Smuggling Suspect

Borderline situations exist where it can be doubtful if all the requirements for *bona fide* medical exposures exist. These can include: self-presentation, some unapproved screening programmes including IHA (Scenario 8 above), unintended and frankly non-medical exposures, some of which take place in medical settings. These can include exposures undertaken for reasons such as: security, smuggling, crime prevention, migration control, aspects of child protection and exposures arising from litigation (EC 2011).

How do these exposures and medical exposures differ? Medical exposures are justified in accord with ICRP's three level process, when it can be confidently asserted that the benefit to the patient will outweigh the associated risks. It is also implicitly assumed that the exposures are conducted within the framework and practices of medicine, which require commitments to consent, confidentiality and respect for the dignity/autonomy of the individual. These commitments are unlikely to be replicated outside of the governance framework for medicine, for example, in a migration centre, customs facility or prison (Malone 2011b, 2016).

Important issues arise in the subset of human exposures that appear like medical examinations but are not conducted for the benefit of the person involved (i.e. they are not medical procedures in the normal sense – for example, drug searches, weapon searches, screening of migrants). Scenario 9 deals with Ms Auburn, age 28 years, who arrives at the airport after a long-haul flight. She is behaving nervously, and sniffer dogs alert the authorities to check her baggage. They find a small amount of cannabis in one of her bags, and she continues to behave suspiciously.

After some deliberation, the authorities decide that she may be a drug mule. The customs officer in charge is also concerned about the serious consequences of a puncture in the drug packaging in the intestines of the young woman. Hence an abdominal/pelvic CT scan is requested from a local hospital with which the customs service has a contract to provide scans in such circumstances. The suspect appears otherwise healthy. She is not appraised of the radiation exposure or any other risks associated with the scan and permission is not sought. She strenuously objects to the procedure but is advised that she will be held until it is performed. Eventually, she consents so that she can go home, as she knows she is innocent. The scan is then performed promptly, competently and with optimised settings.

The report is made available to the customs service. It shows no sign of concealed drugs in body cavities, but also shows Ms Auburn to be pregnant. The customs officer promptly advises her of both findings. Ms Auburn had thought she might be pregnant but was not sure, and the pregnancy issue was not mentioned to her prior to the scan. She is distressed by the situation, but the customs officer is unable to offer her any advice on the pregnancy or the possible radiation damage to her foetus.

The hospital staff assumed the scan was justified by the customs service, which, in turn, assumed the hospital, dealt with this issue. In the ICRP system, this scan doesn't have value to Ms Auburn as a medical procedure, and hence is not medically justified. The requirements for consent and confidentiality are also dispensed within the process described here. In the circumstances, the hospital should have robust authorisation processes to mandate such departures from normal practices and to ensure public confidence cannot be undermined.

TABLE 4.9 Scenario 9: Ms Auburn, Drug Smuggling Suspect

- Ms Auburn (28) at Airport. Sniffer finds small amount of drugs in hand luggage. Suspicion she may be mule.
- Refuses permission for CT scan. Customs officer insists, and hospital does the scan. She is pregnant and no drugs.
- Issues re-scan, around Justification, benefit to individual, confidentiality, consent, and governance arrangements.
- Hospital-based scan and lack of clarity on who justifies.
- Risk of undermining privileges of medical procedures.

Dignity Autonomy	**Beneficence, Non-Maleficence**	**Justice**	**Prudence Precaution**	**Honesty Transparency**
(y)	(–)	(Y)	(y)	(–)
(N)	(N)	(N)	(N)	(N)

Likewise, the customs service must have an open and transparent protocol detailing how such scan requests are justified and who the authorising officer is. On investigation it was found that the hospital and the customs service had negotiated a detailed protocol covering all these issues, but the staff on the ground had not been made aware of it. In some countries, a judge must underwrite the request for a non-medical radiological examination, as the risks and benefits involved are essentially social rather than medical (Malone 2011b).

All the headings in the table at the bottom of Table 4.9 have an **(N)** entry especially when assessed from the perspective of a medical exposure. Ms Auburn's dignity/autonomy are put aside in the performance of the examination and perhaps even more clearly in disclosing the scan results to third parties, particularly about her pregnancy. A justification process that falls between stools and fails to establish the subject is pregnant before the examination is performed compounds this. There are also problems under the headings of non-maleficence, justice, prudence, and honesty. On the positive side, there is a social benefit, which is shown as a **(Y)** under justice accompanying the **(N)** already scored there on Ms Auburn's behalf. The customs officers treat her as well as the situation allows, giving small **(y)'s** under dignity/autonomy and prudence. The latter is awarded in respect of the customs officer's concerns in connection with the potentially lethal risk of burst packaging.

From the above, this scenario is a frankly non-medical exposure. Ensuring good justification and compliance with the pragmatic value set needs much additional consideration when the assumptions of the medical model do not apply. This is the case especially when the exposure takes place in a medical facility and is performed by healthcare employees. For example, to ensure good justification of non-medical exposures, considered answers or protocols dealing with the following questions must be available to staff at the operational level:

- What is the justification process to be followed?
- What are the grounds for justification of such exposures?
- Is the justification process grounded in a set of ethical values?
- Is the justification individual or collective or both?
- Is there a statutory basis for it?
- Which professionals are responsible? For example, is it radiologists? (*who may not be trained in public health, security, childcare issues etc.*); or judges?/customs officers? social workers (*who may not be trained in medical issues and radiological risk benefit analysis*).

Localised weak non-medical justification will, if it persists in medical facilities, inevitably add to the accountability and transparency issues arising from an already significant level of unjustified medical exposures. The pragmatic value set helps flag the issues involved more clearly. Definitive solutions to these dilemmas may require thinking outside the box, to avoid the twin problems Naom Chomsky identifies in major crises in many areas: under-pricing risk and ignoring the impact of actions on external groups (Chomsky 2008; Polychroniou 2016).

4.4.3 Scenario 10: Failed Equipment

One of the main functional rooms in Black Tulip Hospital Interventional Radiology Suite had a breakdown that required replacement of the X-ray tube. The equipment supplier sources a new part, plans to have it delivered three days later, and to install it immediately thereafter. Once installation is complete, the medical physicist, Dr Russet, is contacted to acceptance test the equipment following a major repair and certify it as safe for clinical use (EC 2012). Dr Russet is committed to acceptance testing a CT scanner elsewhere and advises he will be available in four days. The Head of Interventional Radiology, Dr Cinnamon, is reassured by the company engineer that it would be ok to take patients immediately and procedures are undertaken on 35 patients prior to the arrival of Dr Russet, who acceptance tests the machine out of normal working hours so that it will be available for use the next day. Dr Russet finds a filter appears to be missing and the automatic exposure protocols/controls are functioning to give patient doses in the range of 2–10 times higher than was the case before the tube was changed. In consequence, the 35 patients received significantly higher doses than necessary. Dr Cinnamon is very upset, but decides the patients needn't be told, as the information may unduly worry them. Dr Russet advises him that there is an obligation to let the patients know what happened, and furthermore, there is a duty on the hospital to inform the regulator. Dr Cinnamon decides to do neither, and to allow the matter be adjudicated on at the radiation protection committee, scheduled to meet in three months. He also summons the engineer and his manager from the company. It emerges that the engineer is inexperienced and was assigned in response to Dr Cinnamon's insistence that the tube replacement be dealt with immediately.

Assessed against the pragmatic value set, this scenario scores a **(N)** under all five headings (Table 4.10). Using equipment that has not been verified to be safe fails to respect the dignity/autonomy of each patient; the increase in dose is unjust in imposing a larger fraction of the burden of risk on the patients examined with the untested equipment. Likewise, it is imprudent and, finally, it is dishonest in not explaining what is happening/has happened to the patients. The patients should have been advised of the use of untested equipment before their examinations were performed. Some of the negativity is counterbalanced by a **(y)** under

TABLE 4.10 Scenario 10: Failed Equipment

- Black Tulip Hospital Interventional Radiology Suite has a tube failure. Urgent replacement by the company three days later.
- Physicist, Dr Russet, contacted to test if system is safe.
- Dr Russet is commissioning a CT elsewhere, and advises he will be available in four days.
- Dr Cinnamon, Head of Interventional, is reassured by the company engineer, and decides to take patients immediately.
- Dr Russet tests the equipment, it appears a filter is missing and exposure protocols are incorrect, giving doses ×2 to ×10 high.
- 35 patients receive the high doses. Dr Carmine decides they shouldn't be told and refers problem to RP committee.

Dignity Autonomy	Beneficence, Non-Maleficence	Justice	Prudence Precaution	Honesty Transparency
(y)	(–)	(–)	(–)	(–)
(N)	(N)	(N)	(N)	(N)

dignity and autonomy arising from Dr Cinnamon's sense of urgency to have the patients' examinations expedited in a timely way.

The scenario described is not an uncommon one, in that urgent major repairs are often required and the medical physicist/radiation protection expert (RPE) may not always be immediately informed or available. In this case, the medical physicist could have been advised that he would be required sometime in the coming days immediately after the tube failed. This would have allowed some time to rearrange, or to reschedule work, or to seek additional assistance. The dose increase after a major service is not common, but does occur frequently enough for rigorous testing to be advised when major equipment upgrades, servicing or component replacement takes place. Likewise, lesser failures, for example in image quality and other aspects of equipment performance, can and do occur and may well also be unacceptable in equipment used on patients (EC 2012).

4.5 CONCLUSIONS

The pragmatic value set provides an effective roadmap in the evaluation of the scenarios described here. It helps reach decisions that are likely to be socially acceptable and respected. The 10 scenarios explored illustrate how utilisation of the pragmatic set can aid and add to the decision-making framework used for radiation protection in radiology. The values supplement the ICRP principles of justification, optimisation and dose limitation, and complement them considerably in aiding decision-making in socially sensitive areas. With the publication of the ICRP general advice on ethics in radiation protection and pending its report on ethics in the medical area, the value set provides an intuitively clear and credible basis for assessing events, the behaviour of health professionals and institutional protocols in medical radiological imaging (ICRP 2018).

The scenarios identify areas worthy of attention in current practice. Perhaps among the more important are the lack of emphasis on the autonomy and dignity of the individual as well as a lack of emphasis on prudence in some professionals and professional organisations. This, illustrated in several scenarios, is an area in need of considered attention in radiation protection in medicine. Both values are accepted in the wider community and have received significant endorsement from high-level sources including the UN and the World Medical Association (Chapters 2 and 3). On the other hand, excessive reassurance, which is a common response to the medical uncertainties that accompany radiology, can breech honesty, and is not appropriate in the face of the real, if small, probability of serious harm. Such reassurances ultimately damage credibility. Furthermore, patients constantly encounter and cope with much larger uncertainties in other aspects of the medical interventions they experience.

Other values from the pragmatic set, particularly non-maleficence/beneficence, come within the ambit of what was traditionally required of optimisation, and hence are, perhaps, not quite so challenging. There is a good history of successful implementation in the area, although much more is possible. Likewise, in managing health services, considerations of justice often arise and, while much remains to be done, there is a good level of awareness of it as an issue in radiology. Finally, the culture of radiation protection in medicine is challenged by the demands of honesty. Messages to the public, framed within that culture, are often paternalistic, evasive and are sometimes viewed with suspicion. This is clearly illustrated in some of the scenarios. In the final analysis, an enduring solution will require that the patients and public be given the truth in clear language they can relate to, and be involved in decisions, particularly where uncertainty prevails (Chapters 6 and 7).

Some human exposures pose additional ethics and regulatory problems, particularly exposure of pregnant or potentially pregnant persons, asymptomatic persons, and non-medical human exposures undertaken in medical settings. As demonstrated, the pragmatic value set provides a broad perspective for assessing these, particularly prudence and dignity. This may be even further augmented by considering the additional values discussed in Chapter 6 and/or the more general considerations in Chapter 7. Finally, the help and advice of those involved in the legal and social care professions will bring additional important new perspectives to this area (Ricomet 2018). From these, we can expect many new insights on how best to practice radiation protection of patients.

CHAPTER 5

Ethics Analysis of Radiotherapy Scenarios

5.1 INTRODUCTION

This chapter is dedicated to the application of the pragmatic set in various common situations encountered in radiotherapy (RT). The scenarios do not discuss conduct traditionally described as 'unethical' or 'illegal'. Lord Kilbrandon observes in the inaugural issue of *Journal of Medical Ethics* that 'the most fascinating aspect of medical ethics in the broad sense is that controversies within its boundaries can be most acute just where knowledge, skill and enthusiasm are at their most advanced' (Kilbrandon 1975) as it is in the aspect of the following scenarios. Today's medicine benefits from a multi-disciplinary approach and patient care no longer rests in the hands of the treating physician alone. In radiotherapy, the duty of care once considered as appropriate solely to the doctor is now shared by nurses, social workers, dieticians, clinical psychologists, radiation therapists and medical physicists. The scenarios were created to reflect the cross-pollination of values amongst the different disciplines. The pragmatic set discussed in Chapters 2 and 3, namely dignity and autonomy of the individual, non-maleficence and beneficence, justice, prudence and honesty, provide a moral framework for examining attitudes described in the scenarios.

One of the challenges in the radiation protection of the patient in radiotherapy is the paradox of an increased likelihood for future exposure if the outcome of the treatment is favorable. The disease free interval and the overall survival of cancer patients has greatly improved over the last few decades due to technological improvements in the delivery of radiation and the recruitment of other monotherapies to aid where radiotherapy alone was suboptimal. Examples include the introduction of intensity modulated radiotherapy (IMRT) techniques, which has increased the overall survival of Hodgkin lymphoma patients (Parikh et al. 2016), and the addition of hormone therapy to radiotherapy which has increased survival of prostate cancer patients (Kumar et al. 2006; Mottet et al. 2012). With patients living longer, the possibility for manifestation of metastatic disease or another primary cancer increases. Follow-up care plans almost always include imaging scans at regular intervals and detection of further disease is likely to result in administration of additional radiotherapy.

The flip side is the chance for developing secondary cancer due to the radiation dose received for the treatment of the first. Adolescent girls who have received mediastinal radiation therapy for Hodgkin's disease are 20 times more likely to develop breast cancer than the general population (Bhatia et al. 1996). Women who have had radiotherapy as part of their treatment for breast cancer are at a higher risk of developing lung cancer than those that did not (Grantzau et al. 2014). The choice of the appropriate treatment, knowing that potential harm may be expressed if the treatment is successful and the patient's life is prolonged enough, is undoubtedly among the more frequent ethical dilemmas a radiation oncologist faces. It must be stated that this choice cannot be made on ethical grounds in the absence of up-to-date scientific knowledge of the outcomes of clinical trials, of the development of new techniques and of advances in the fields of chemotherapy, surgery, pathology, and radiology to name a few.

There has been a great reform in the new millennium in the role of the professional in protection of the patient in radiotherapy. Breakthroughs in technology allow us to design and deliver radiation more accurately and treatment planning systems have adopted dose calculation algorithms that can more accurately predict the dose received by heterogeneous tissue in the patient. These in turn result in improved tumour control probability and reduced normal tissue complications. These changes in the social environment in which radiotherapy is practiced require skills beyond scientific and technical skills. This reform requires us to revisit the values on which treatment decisions are based as the medical and scientific principles alone are proven inadequate. Autonomy is challenged by the complexity of the treatments and the uncertainty of the outcome, combined with the changing patient demographic and their increased desire to understand and be involved in the choices required to manage their disease. Observing beneficence and non-maleficence in cancer care is now a more complex task than simply increasing survival or providing palliation, and demands that we consider the overall impact on the patient's quality of life. Justice must balance availability and distribution of resources with the benefits in the use of radiotherapy for each patient. Honesty invites us to expose weaknesses and uncertainties in our practice in the era when any error or transgression can be broadcasted worldwide in a manner of minutes. The healthcare worker in radiotherapy must have the wisdom to *recognise and follow the most reasonable course of action, even when full knowledge of its consequences is not available.* That wisdom which harmonizes the values of medical ethics radiation protection and scientific vigor is reflected in the principle of prudence as introduced in Chapter 2. Prudence, alongside with transparency, are emphasized in the scenarios discussed in this chapter. Prudence must drive decisions regarding a patient's treatment in a realm of ever changing technology, a plethora of scientific data and clinical studies, and the inherent uncertainty of radiobiology. It is a value of particular significance in radiotherapy and it is likely that many radiotherapy professionals are unfamiliar with the term even though we will see that the concept is commonly encountered in daily practice.

While this chapter does not provide any prescriptive advice on how a medical professional ought to apply these values in their practice, it is hoped that the analysis of the scenarios will stimulate ethical reflection and discussion among all professionals involved in the protection of the patient from radiation.

5.2 ASPECTS OF SCENARIOS AND METHODOLOGY

The analysis of fictional scenarios in this chapter follows the methodology introduced in Chapter 4. The ten scenarios are analysed in terms of the pragmatic value set established in Chapters 2 and 3.

In the first scenario, a physician wishes to treat her patients differently to what the departmental protocol prescribes. Off-protocol treatments are not in themselves unethical; their use, however, must be justified and prudence must be exercised. The second scenario introduces us to a patient who interprets his right to healthcare as encompassing his wishes regarding location and method of treatment. The rise in patients' expectations is of particular ethical interest in healthcare as the principles of autonomy and justice come under direct scrutiny. Justice stands out amongst the other values as it invites one to consider those not directly involved in the interaction. For physicians to act justly, they must not only consider what is best for the patient before them, but also how any decision may affect other patients (also refer to Chapter 7). The constant balancing of resources due to the finite availability of treatment equipment, time, and financial capital calls for continuous evaluation of availability of alternative options.

The same scenario also brings to light another question regarding autonomy and the involvement of the patient in the decision making of his care. Much ink has been spilled in recent decades on the issue of consent. Challenges regarding its content, its power, the processes involved and its deployment are examined further in the third and fourth scenarios. The highly technological nature of radiation therapy poses a particular challenge in explaining the details and options for treatments to the patient, and hence to how autonomy and consent are interpreted.

Alongside consent, the complications of pregnancy and mental capacity are weaved into the third scenario. The lack of general legal and ethical consensus surrounding the rights of the mother and with whom the duties of the healthcare staff lie, are reflected in the difficulty with which the physician and the radiation therapists in the fourth scenario make any judgment calls. Cases of pregnant patients are particularly formidable, and therefore legal statutes have attempted to aid medical decision. In England and the United States, the foetus is not regarded as an independent person until it is born, which gives the mother, and the State, the capability to act to the best interests of the mother. In other jurisdictions, such as Ireland, the provisions of the Constitution (up to 25 May 2018) obliged the State to protect the equal right to life of the unborn child, making the right of a woman and the interest of her unborn child difficult to balance. This scenario of a pregnant patient is an example where the law is not a good guide for ethical deliberation and decisions must be illuminated by a different light.

The fifth scenario opens the door to the dilemmas surrounding clinical trials and the juxtaposition of beneficence, non-maleficence, and justice. Clinical equipoise and the principle of justice are further expanded in the sixth scenario. Social and economic constraints create a problem for managing limited resources aiming for a fair outcome using a fair process.

The pragmatic and political concerns that may have an impact on the decision for type and complexity of treatment are explored in the next two scenarios. In Scenario 7, we meet a physician contemplating the use of advanced treatment for a patient. Radiotherapy is a discipline characterised by rapid evolution both in the techniques available for treatment and in its use in combination with other treatment modalities, such as immunotherapy and/or surgery. When new techniques are available but robust data to establish their status against competing existing techniques are absent, medical professionals must rely on their own experience and judgment. As discussed in Chapters 2, 3, and 7, when uncertainty is part of the drama, prudence and precaution must take the centre stage. The eighth scenario describes a scene familiar to many physicians, where a dying patient expresses a symptom and the physician must advise on whether it should be addressed or is better left alone. Precaution in such cases may come in direct conflict with beneficence/non-maleficence and with justice. To take action in a healthcare setting is far easier than to do nothing as it satisfies an altruistic or charitable tendency. However, end-of-life scenarios are ones in which the involvement of the patient and their support network in any decision making is of paramount importance.

In Scenario 9, a physicist is put under pressure to make a new treatment technique available for clinical use in an environment of limited resources. When new evidence is made available for the benefit of a technique or pharmaceutical agent, healthcare providers have an obligation under beneficence to, where possible, make it accessible to their patients. However, prudence and precaution must also be observed in such situations.

The final scenario contemplates prudence, honesty, and transparency. A human error occurs which transpires into the mistreatment of a patient. Minor errors occur daily in every radiotherapy facility, most of them with no significant impact on the outcome of the treatment. Some significant errors, however, can and do occur and cause serious harm to the patient. Where precaution is the prevalent value in error prevention, respect for the patient's dignity and transparency are commonly implicated when errors occur.

The evaluation of compliance to the pragmatic value set of the professionals introduced in each case is the personal judgment of the authors. Compliance with a value is indicated as being strong **(Y)**, weak **(y)** or neutral **(–)**. Likewise, Non-compliance is indicated as strong **(N)**, weak **(n)** or neutral **(–)**. Some aspects of scenarios demonstrate compliance with a value when considered from one perspective, and non-compliance when considered from another. Thus, it is possible to score both **(Y/y)** and **(N/n)** for the same value. The methodology is discussed at length in Chapter 4 and the reader is urged to review it before delving into the cases presented and analysed in this chapter. The scenarios in this book are not exhaustively analysed. There are many layers to ethical contemplation of any complex medical situation. The analysis presented here serves as a demonstration of how the pragmatic value set can be used, while simultaneously validating its suitability in radiotherapy.

5.3 SCENARIOS

5.3.1 Scenario 1: Dr Loeen, Imaging Frequency

At Spring Clinic, all radiotherapy patients undergo CT scanning to produce image sets for creating a treatment plan and calculating the estimated dose to the target and surrounding tissues and organs. This scan is in addition to all previous diagnostic scans. The scan length often extends further than the already identified region of disease in order to capture the patient's position and dose information to allow them be replicated during treatment. Patients are consented for this procedure as part of their radiotherapy.

To verify dose and positioning reproducibility, Spring Clinic scans patients weekly during the course of treatment. The radiation risk from the weekly scans has not been studied, but the reduction in some treatment related toxicities that can be achieved from implementing such imaging strategies is documented in international journals. The Clinic issued formally approved protocols on the frequency and settings to be used in these weekly scans.

Such practices are common in most radiotherapy centres and are an example of additional dose from imaging specific to radiotherapy patients (i.e. a patient receiving only chemotherapy as part of the their treatment will not be exposed to this radiation). The additional dose, however, is justified by its presumed benefit in informing an optimized and individualized plan for radiotherapy treatment. Evaluating this practice against the pragmatic set, we would score autonomy with a capital **(Y)** as the patients are consented, and beneficence with both a small (**n**) and a capital (**Y**). However, let's examine the practice of an individual radiation oncologist, Dr Loeen.

Dr Loeen recently joined Spring Clinic as a staff radiation oncologist. She requests that all her patients be scanned daily while on treatment. Even though she cannot show evidence of benefit to the patient, she argues 'If weekly imaging is good for the patient, daily imaging can only

TABLE 5.1 Scenario 1: Dr Loeen, Imaging Frequency

- Radiation Oncologist Dr Loeen, new member of staff at Spring Clinic.
- Established imaging frequency and correction strategy for all radiotherapy patients.
- She requests that her patients follow a more rigorous imaging strategy with daily setup correction without proof of benefit.
- Practice is refused by the medical physicist.
- She accuses him of suboptimal care.

Dignity Autonomy	Beneficence, Non-Maleficence	Justice	Prudence Precaution	Honesty Transparency
(y)	(y)	–	–	–
–	(N)	–	(N)	(N)

be better'. Dr Loeen's request is made with her best intentions for the well being of her patients. However, it will cause a fivefold increase in the number of scans. The additional risk with no proven benefit scores an **(N)** for both non-maleficence and prudence. Mr Brickfielder, the medical physicist in the department who has calculated the increase in dose to the patient from daily scanning, understands that it is his professional responsibility to question Dr Loeen's instructions. He insists that any change in practice unsupported by evidence should be initiated under the auspices of the clinical trials unit with ethics approval. To initiate such a study is time consuming and Dr Loeen tells Mr Brickfielder to 'Stay out of it' and that 'Her patients are not going to receive suboptimal treatment because of bureaucracy'. The choice to ignore expert advice and follow established pathways to ensure the safety of the patient scores a capital **(N)** for honesty. As her proposed practice is justified by her experience and may be proven beneficial to the patient if properly studied, and as she will consent the patient for the additional exposure, her practice scores a small **(y)** for beneficence and for autonomy (Table 5.1).

5.3.2 Scenario 2: Mr Wilson, A Patient's Right to Choose His Treatment

Mr Wilson is diagnosed with squamous cell carcinoma (SCC) on the left alar wall (nose) of 2 cm diameter extending towards the left lower eyelid. The disease extends to a few millimeters below the skin surface. Dr Boysenberry, the radiation oncologist looking after Mr Wilson, has offered him treatment on an orthovoltage X-ray machine at Asclepius Clinic 16 km away from Mr Wilson's home. Mr Wilson becomes aware of a different clinic 2 km from his house, which does not have an orthovoltage treatment unit but can treat conditions like his with an electron beam.

He therefore requests to have his treatment there as it is more convenient. Dr Boysenberry explains both techniques emphasizing the benefits of orthovoltage in Mr Wilson's case. He also discusses the process and the waiting time that would be required to treat Mr Wilson at the near-by clinic to allow Mr Wilson to understand why the specific treatment is offered to him. This scores Dr Boysenberry's actions a capital (**Y**) under autonomy and transparency.

However, Mr Wilson refuses to attend treatment at Asclepius Clinic and insists on receiving his treatment at the nearest clinic. Mr Wilson argues that he is acting on his right to determine what should be done to his own body. He has what he considers to be adequate information at hand to choose the facility he wishes to attend. The question we must ask is whether Dr Boysenberry must accommodate his patient's desires in order to honour his dignity and autonomy. However, the physician must also consider the value of justice when selecting a treatment. Mr Wilson is indifferent to the fact that in order to receive treatment at the near-by clinic with an electron beam, a complex physical setup must be created which is not necessary for treatment on the orthovoltage unit. He is also indifferent to the fact that the treatment must be delivered on a congested linear accelerator that is also capable of treating patients who cannot be treated on an orthovoltage unit.

Dr Boysenberry decides to only offer treatment to Mr Wilson at Asclepius Clinic and provides him with patient transport options to facilitate his treatment. He does not view this as disrespecting his patient's autonomy. On the consent form, Mr Wilson is asked to agree to receiving radiotherapy for his cancer and accept the toxicity risks of the treatment as discussed with him. Dr Boysenberry takes advantage of the ambiguity in the object of the consent that does not include specifics of the type of radiotherapy, location, or duration of treatment.

As the treatments at each clinic are equally effective, not accommodating Mr Wilson's request does not compromise his treatment and therefore still scores a (**Y**) for the value of beneficence. Dr Boysenberry's decision to offer treatment on the equipment that he knows to be more appropriate for the type of cancer Mr Wilson has, also scores a (**Y**) under prudence. In addition, a further (**Y**) is added under justice. If Dr Boysenberry refused Mr Wilson treatment at the near-by clinic in the name of justice without offering an alternative or an explanation, which would have been equivalent to harming him and disrespecting his dignity and therefore could not be ethically justified. In light of the option at Asclepius Clinic, albeit

TABLE 5.2 Scenario 2: Mr Wilson, A Patient's Right to Choose

- Mr Wilson is offered treatment at Asclepius Clinic. He becomes aware of an equivalent treatment available at a near-by clinic that is more convenient to him.
- Both treatment techniques are explained and consent for treatment is obtained.
- Due to the demand of the treatment unit at the near-by clinic, Mr Wilson is only offered the option of having treatment at Asclepius Clinic.

Dignity Autonomy	**Beneficence, Non-Maleficence**	**Justice**	**Prudence Precaution**	**Honesty Transparency**
(Y)	(Y)	(Y)	(Y)	(Y)
(n)	–	–	–	–

of some inconvenience to Mr Wilson, Dr Boysenberry is respecting both values of justice and the beneficence (Table 5.2).

5.3.3 Scenario 3: Dr Celadon, Consent

Ms Ruddick recently completed irradiation treatment for lymphoma to her supraclavicular fossa followed by chemotherapy at Apollonean Clinic under the care of Dr Celadon, a radiation oncologist. Ms Ruddick is the mother of two toddlers. She has been struggling with borderline personality disorder for over a decade. She was advised not to get pregnant while receiving chemotherapy. Ms Ruddick is now presenting to Dr Celadon with recurring aggressive disease and is 8 weeks pregnant. Dr Celadon explains her condition and the prognosis of 6 months survival her condition carries if left untreated. Ms Ruddick is agreeable to discussing treatment options for her disease. She is offered information about termination of pregnancy, which she categorically refuses. During consultation with Dr Celadon she demonstrates little understanding and, as she is fearful of the risks of chemotherapy to her unborn child, refuses all systemic treatment. Dr Celadon discusses radiotherapy options with Ms Ruddick, scoring a **(Y)** under autonomy and transparency. However, he does not enlist the assistance of a psychologist, social services, or engage with the patient's family to ensure that she fully comprehends her situation. This also adds a small **(n)** under autonomy. He explains to Ms Ruddick that to not avail of any treatment is almost certainly a death sentence for both her and her unborn child. Large field radiotherapy, which may improve her prognosis, will expose the fetus to levels of radiation associated with miscarriage and growth retardation. As he appreciates the limited capacity of his patient to understand and weigh her options, he offers a third option, radiotherapy to the mediastinum alone (limited field), which will reduce the dose to

TABLE 5.3 Scenario 3: Dr Celadon, Issues of Consent

- Ms Ruddick, borderline personality disorder, presents with recurring disease.
- Ms Ruddick is also 8 weeks pregnant at the time of presentation.
- Dr Celadon, a radiation oncologist, explains her options and the impact of each to both her and her unborn child.
- The patient demonstrates poor understanding of both her condition and her options. Dr Celadon offers her the treatment he knows is best, and Ms Ruddick consents.
- The treatment chosen will prolong Ms Ruddick's life enough to come to full term and carry some risk to foetal development.

Dignity Autonomy	Beneficence, Non-Maleficence	Justice	Prudence Precaution	Honesty Transparency
(y)	(y)	–	(Y)	(Y)
(N)	(n)	–	–	–

the uterus and could prolong her life enough to bring her to term. The later also increases the risk for growth retardation and major organ malformations, but not as much as large field treatment. The balancing between the options score Dr Celadon a (**Y**) under prudence. Since there is both harm and good resulting in the proposed treatment, both a (**y**) and an (**n**) are added to Table 5.3 under beneficence/non-maleficence. Dr Celadon strongly recommends the latter and asks Ms Ruddick to consent to it during her consultation so as to not delay the beginning of treatment. The behaviour of Dr Celadon and the hastiness with which he is imparting on his patient for her decision, albeit with the patient's best interests at heart, is highly paternalistic and changes the (**n**) to a capital (**N**) under autonomy.

5.3.4 Scenario 4: Mr Catreuse, Who Bears the Responsibility?

Ms Ruddick, now 10 weeks pregnant, attends an appointment where a CT scan necessary for treatment planning is to be performed. Mr Catreuse, the lead CT radiographer at Apollonean Clinic, ensures that the scan length does not include the uterus and hence does not pose any risk to the fetus. He explains this to Ms Ruddick who agrees to have the CT scan. Just before scanning, Ms Ruddick requests a pause and expresses concern and distress about the wellness of her unborn child. She acts erratically and asks if she is 'going to die'. Mr Catreuse reassures her that this scan does not put her child at any risk and calls in Dr Celadon, the radiation oncologist looking after Ms Ruddick. Dr Celadon reminds Ms Ruddick of the discussions they have already had about her prognosis and plan for treatment and instructs the radiographers to continue with the scan. Mr Catreuse and his team of radiographers act on the written consent given previously by Ms Ruddick and the current verbal directive of Dr Celadon without any knowledge of the patient's character or psychological condition. However, her behaviour during the scan demonstrates unclear understanding of both her medical condition and the risks to herself and her unborn child from the scan and the treatment.

The radiographers in this case assume that the previous consent is an expression of the autonomous wishes of the patient acquired at a time when the patient was competent to express her desire. They are unmoved by the paternalistic behaviour of her physician when Ms Ruddick is demonstrating a lack of understanding of her condition or the procedure she is undergoing. The staff actions, while based on a previous consent, score a capital **(N)** under autonomy/dignity. An approach more respectful of the patient's autonomy in a situation such as this might be to postpone the scan, giving the patient additional time to reconsider her desires and decisions. Mr Catreuse's caution to limit the scan length and exclude the uterus scores a capital **(Y)** both for non-maleficence and prudence. The staff's ignorance of the history of the patient they are scanning and their unquestioned compliance with the doctor's instructions score a capital **(N)** under honesty and transparency, while a small **(y)** can be justified by their engagement in explaining to the patient that the current procedure does not pose any risk to her (Table 5.4).

This case raises the topic of sharing the moral responsibility of patient care with the other members of the health provider complex. In radiotherapy, as in other disciplines, it is easy to compartmentalise procedures and assign responsibility to staff only for the part in which they are directly involved. Traditionally, the comprehensive, moral responsibility for patient care remained with the physician, and this

TABLE 5.4 Scenario 5.4: Mr Catreuse, Who Bears the Responsibility?

- Mr Catreuse is the radiographer in charge of CT scanning.
- Ms Ruddick, a pregnant patient with a history of bipolar disorder, is having a CT scan for radiotherapy treatment planning purposes.
- Mr Catreuse limits the scan length to protect the uterus.
- Ms Ruddick demonstrates unclear understanding of the purpose and the risk of the procedure.
- Mr Catreuse involves the radiation oncologist who convinces the patient to comply with the scan.

Dignity Autonomy	Beneficence, Non-Maleficence	Justice	Prudence Precaution	Honesty Transparency
(y)	(Y)	–	(Y)	(y)
(N)	–	–	–	(N)

drives the CT staff of Apollonean Clinic to contact the radiation oncologist when their patient exhibits unusual behaviour. This model of care, absolving all but the physician from the duties of ethical (and often legal) deliberation and patient care, is conceptually problematic as the physician is neither omnipresent nor omniscient. When part of the patient's care is entrusted to specialised or expert staff, the care team is given medical custodianship of that patient and with it the fiduciary duty to act in the patient's best interest. To uphold the values of dignity, autonomy, and honesty, the care team must be responsible not only for the procedure and the task assigned to them, but with the comprehensive well being of the patient while in their care.

5.3.5 Scenario 5: Dr Oxley, Clinical Trials

Eligible cancer patients at Hygeia Clinic are offered the option to participate in a clinical trial led by Dr Oxley. The trial is investigating the potential benefit of Intensity Modulated Radiation Therapy (IMRT; a technique where hundreds of thin beams of radiation are used to accurately deliver the prescribed dose to the target and reduce the dose to other organs) against 3D conformal treatment (3D-CRT; a technique where typically 3–5 beams are pointed to the target) in reducing toxicity. Eligibility for the trial is defined in the protocol and excludes patients in specified categories, including 'those with any condition that may compromise the outcome of the study'. The radiation oncologist decides the latter criterion at the patient's first visit. Hygeia Clinic is a satellite of a large academic hospital, the Ethics Committee of which has approved the trial.

For over a decade, patients at Hygeia have been planned and treated with 3D-CRT with outcomes matching those reported by other radiotherapy centres. If the treatment plan fails to meet published dose specifications to the non-affected organs (known as 'organs at risk') the treatment is re-planned using IMRT. This practice obeys the value of prudence/precaution as procedures are in place to potentially minimise harm to the patient. While IMRT allows for more precise distribution of the dose to the target area, the volume of distal tissues exposed to the low dose is increased; the effects of the low dose exposure of distal tissues is not known. The trial is designed to show if there is any benefit from the increased precision in the deposition of the prescribed dose to the target and the sparing of adjacent tissue but it is not powered to show the impact of the increased exposure to low dose.

TABLE 5.5 Scenario 5: Dr Oxley, Clinical Trials

- Dr Oxley is conducting the principal investigator in a trial comparing toxicities and outcomes from *3D-CRT v IMRT*
- Patients at the hospital are routinely treated with 3D-CRT unless they fail to meet dose constrains. In that case they are treated with IMRT.
- Mr Feldgrau is Dr Oxley's patient. He is considering participation to the trial believing that he will avail of IMRT.
- Dr Oxley does not explain randomization and urges him to join.

Dignity Autonomy	Beneficence, Non-Maleficence	Justice	Prudence Precaution	Honesty Transparency
(Y)	–	(Y)	(y)	–
(N)	–	–	(n)	(N)

The planning and treatment approach will be different for patients enrolled in the trial. Due to randomization, patients who can meet the dose specification on 3D-CRT can still be randomized to IMRT and conversely, patients not meeting these could still be treated with 3D-CRT. The trial removes the precautionary measure put in place by Hygeia Clinic and scores a (**n**) under that value on Table 5.5. The principle of equipose is not violated as there is no scientific proof to support this measure, and therefore a (**y**) is also added under prudence/precaution. Mr Feldgrau, who was recently diagnosed with a disease of poor prognosis, attends Dr Oxley's clinic and is offered participation in the trial. The risks are explained and an information leaflet is given to him; Mr Feldgrau will be consented for the trial at the time of attendance for his planning CT, 10 days later. By providing the leaflet and the time for Mr Feldgrau to study this information, the medical team is fulfilling its ethical obligation for informed consent. When Mr Feldgrau returns 10 days later, he opts for joining the trial because he thinks that by agreeing to it, Dr Oxley will see him more favourably and will pay increased attention to him during his treatment. Despite his misconceptions, the information leaflet and the time for the patient to make his decision score the medical team a (**Y**) under dignity/autonomy.

At his appointment, Mr Feldgrau asks Dr Oxley whether she thinks this trial 'is good for him'. Dr Oxley points out that while this is a randomized trial, there is a high chance for him to avail of the more advanced technique (IMRT). Dr Oxley tends to present the facts in a favourable light to convince patients to sign up as she is eager to complete the trial and publish the results. Not disclosing her academic gain from their decision to her patients is contrary to honesty and respect for the patient's autonomy and score a (**N**) in both values.

She is not violating the beneficence or non-maleficence principles as the two techniques have not been shown to have a different outcome yet. The risk to Mr Feldgrau is not increased as there is no evidence to indicate that the switch to IMRT is of any added benefit or detriment. His participation however will benefit future patients who will be availing of the knowledge gained from the result of this trial. Therefore, the encouragement for the patients to enroll in the trial is also serving the justice principle scoring a (**Y**).

5.3.6 Scenario 6: Ms Thomson, Patient Prioritisation

The *3D-CRT v IMRT* trial at Hygeia Clinic is completed and reveals reduction in radiation related toxicities when patients are treated with IMRT. The trial shows no survival or local control benefit from the use of IMRT. Hygeia Clinic has 2 linear accelerators capable of IMRT and 2 accelerators capable of 3D-CRT. It also has a policy where head-and-neck cancer patients, due to proven survival benefits of IMRT, are prioritized to the IMRT accelerators. Ms Thompson is the wife of the director of Hygeia Clinic and is seeing Dr Oxley for her radiotherapy treatment. Dr Oxley has been treating patients using the 3D-CRT technique for over a decade. She knows that most of her patients have tolerated 3D-CRT well and their toxicities have been controlled with medication. However, she wants to give Ms Thompson the 'best treatment available' and offers her IMRT. The waiting time for IMRT is 5 weeks for non-head-and-neck cancer patients. The impact on the treatment outcomes of the delayed start is not known, but she speaks to the service manager who is able to postpone the start date of a head-and-neck patient to accommodate Ms Thompson's treatment in 2 weeks time. She has not mentioned the details of one technique over the other to Ms Thompson as she feels that it is unnecessary. She argues 'surgeons don't discuss which scalpel or which navigation system to use with their surgical patients, why should it be different for radiation oncologists? The doctors are the best equipped to make such decisions for their patients'.

Cases such as this are not uncommon in general medical practice, and radiotherapy is no exception. Most people in a position to affect the care given to patients known to them, or are within a couple of degrees of separation from them, are likely to do so. In the case of Dr Oxley, she has the added pressure of the patient being close to her superior. Her position gives her the influence to ask for treatments to be expedited or to offer 'the best treatments available'. The prevailing value in these scenarios is justice. As discussed earlier, justice expects

TABLE 5.6 Scenario 6: Ms Thompson, Patient Prioritisation

- Mr Thompson is the director of Hygeia Clinic. Ms Thompson is to have her radiotherapy under Dr Oxley.
- Department policy is to prioritise H&N patients for IMRT and offer IMRT to other patients only when not meeting dose specifications.
- Dr Oxley offers IMRT to Ms Thompson and arranges to prioritise her over H&N patients.
- Ms Thompson is ignorant of this arrangement.

Dignity Autonomy	Beneficence, Non- Maleficence	Justice	Prudence Precaution	Honesty Transparency
(y)	(Y)	–	(Y)	–
(N)	–	(N)	(n)	(N)

one to evaluate the impact of decisions on those beyond the individuals involved in the transaction. In this case, Ms Thompson, without her knowledge, is negatively affecting the treatment of another patient. This behaviour not only scores a capital (**N**) in justice, but also a small (**n**) in autonomy as the patient is not given the opportunity to participate fully in the management of her treatment. Dr Oxley's comments show her paternalistic approach to care, changing to an (**N**) in autonomy and adding another (**N**) under transparency and honesty. Dr Oxley is acting on what she believes to be in the best interest of Ms Thompson, sparing her both of the toxicities associated with 3D-CRT (albeit manageable with medication) and physical and emotional detriment of the waiting time for treatment on the IMRT linear accelerator. Beneficence and non-maleficence are a strong (**Y**) in this case. Prudence is also marked with a (**Y**), as Dr Oxley's actions offer her patient the treatment which is less likely to cause her severe side-effects. However, it also scores an (**n**) as the effects of the low-dose bath associated with IMRT are not known (Table 5.6).

5.3.7 Scenario 7: Ms Payne, Choice of Treatment Technique

Ms Payne is 82 years of age and was treated for breast cancer 17 years ago. She underwent full treatment and was quite certain she had beaten this disease as all her follow-up scans have retuned negative until now. Her latest MRI showed 3 small lesions in her brain, confirmed by radiologists to be consistent with metastatic disease from breast cancer. She has kept up with her monthly book club meetings and her oil painting and feels extremely well. Upon receipt of the results, she attends Dr Perinone's clinic, a renowned radiation oncologist known for treating neurological cancer. Dr Perinone works at Spectral Clinic, a private clinic equipped with the technology for radiosurgery (SRS), a technique where a very high dose is accurately delivered to small, localised tumours, while sparing the surrounding brain tissue. The evidence supports SRS for patients under 65 years with 3 lesions or fewer. Dr Perinone, due to Ms Payne's actively engaged lifestyle, offers to treat her with SRS and explains the risks and benefits of the treatment. In order to design the treatment plan, Dr Perinone orders a new higher resolution MRI scan. It demonstrates that the 3 known lesions have increased slightly, and two new lesions are seen. These may have been there previously, but not been detected on the lower resolution scan. The evidence for benefits of SRS in patients with more than 4 lesions is weak. The alternative option is radiation to the whole brain (WBRT). Given the new lesions detected, there is a possibility of the presence of more lesions too small to be detected by MRI, which WBRT can be beneficial in treating. However, WBRT has been associated with deterioration of cognitive function. Despite Ms Payne's age and the lack of robust evidence, Dr Perinone discusses the options with her and offers to treat her with SRS to protect her from cognitive deterioration and allow her to continue with her bookclub and painting.

Choice between treatment methods is one of the ethical dilemmas radiation oncologists must face regularly. Studies can inform decisions for the patients that match the characteristics of the participating cohort; however, that is often not the case in practice. Dr Perinone is balancing the scientific evidence with his own prior experience and the specific circumstances of Ms Payne. In this, he is exercising prudence and scores a (**Y**). While SRS will treat the visible lesions (hence a (**Y**) in beneficence), there is a chance that by not giving her WBRT, more lesions will appear in a short interval that will negatively affect her survival. For that, a small (**n**) is added under the same value. Dr Perinone is showing respect for his patient's individuality by protecting the functions that are important to her, and his ongoing engagement with the patient, scores a (**Y**) under dignity, autonomy, honesty and transparency (Table 5.7).

5.3.8 Scenario 8: Dr Alizarin, To Treat or Not To Treat

Ms Johansen is an in-patient in Airmed Hospital. She has been diagnosed with advanced stage cervical cancer, which had also invaded her bladder. She has urinary obstruction due to the cancer in her bladder and has a urinary catheter inserted. Dr Alizarin is called to consult on her treatment after a nurse notices excessive blood in Ms Johansen's urine bag. Ms Johansen's health has deteriorated since she was admitted a week

TABLE 5.7 Scenario 7: Ms Payne, Choice of Treatment Technique

- Ms Payne, aged 82, is a breast cancer survivor, an active painter and an avid book reader.
- She now presents with three intracranial metastases and is offered stereotactic radiosurgery (SRS) by Dr Perinone, a radiation oncologist.
- Upon further investigation, two additional lesions are detected which put to question the benefit of SRS.
- Dr Perinone offers her SRS over whole brain RT to protect her from potential damage to her cognitive function.

Dignity Autonomy	Beneficence, Non-Maleficence	Justice	Prudence Precaution	Honesty Transparency
(Y)	(Y)	–	(Y)	(Y)
–	(n)	–	–	–

earlier and she is now on morphine for pain control. She is aware of the change in colour of the fluid in her urine bag, but has not expressed any concern or discomfort associated with it. Dr Alizarin, alerted by the hematuria (blood in the urine), prescribes radiotherapy to help stop the bleeding. She is aware that haematuria is most likely due to the presence of cancer in Ms Johansen's bladder, and based on her examination she expects that she only has a few weeks to live. She nonetheless decides to veer on the safe side and prescribe treatment for it. As she walks away, she is skeptical about whether the haematuria was bothering her more than it was bothering Ms Johansen and if she should have just left it alone.

This scenario raises the question of why a doctor chooses to treat and when is that the appropriate choice. Dr Alizarin's thoughts as she walks away show her awareness of this dilemma. Another physician could have seen the blood in the urine and not reacted to it as Ms Johansen is unlikely to live long enough to see the benefit from the treatment. In either case, the physician scores a **(N)** under dignity and autonomy since she did not ask the patient directly if the symptom is causing her discomfort or distress and offer information to help her choose if she wishes to have it addressed. Dr Alazarin scores a **(Y)** under precaution as she is taking action in the case the patient outlives her expectations. She, however, scores an **(N)** under transparency as she does not share her uncertainty with the patient or with any of her colleagues. The treatment is almost certain to increase Ms Johansen's fatigue and cause her additional discomfort, scoring an **(n)** under non-maleficence. On the contrary, it is not expected to increase her life expectancy; however if she is alive long enough she will see the benefits of the treatment in

TABLE 5.8 Scenario 8: Dr Alizarin, To Treat or Not to Treat

- Ms Johansen is an in-patient with cervical cancer invading bladder under the care of Dr Alizarin, a radiation oncologist.
- Dr Alizarin is notified of the presence of blood in Ms Johansen's urine bag.
- Dr Alizarin orders radiotherapy for Ms Johansen to address the haematuria but she is uncertain that Ms Johansen will live long enough to see any benefit.
- Dr Alizarin is contemplating whether she made the right choice.

Dignity Autonomy	Beneficence, Non-Maleficence	Justice	Prudence Precaution	Honesty Transparency
–	(y)	–	(Y)	–
(N)	(n)	–	–	(N)

controlling the bleeding, and therefore scores a (**y**) under beneficence. The value of justice is difficult to address in palliative care scenarios as the use of resources can be justified if the outcome is positive but it can not if the patient does not live long enough to benefit from the treatment (Table 5.8).

5.3.9 Scenario 9: Dr Garnet, Introduction of a New Technique

Dr Garnet is the chief physicist at Sospitas Clinic, a small stand-alone radiotherapy clinic, where the medical team requests the introduction of stereotactic body radiotherapy (SBRT) for their lung patients. This type of radiotherapy targets a small lesion in the lung and treats it with 3–5 fractions of high dose, as opposed to traditional radiotherapy where the dose is delivered in 25–35 smaller fractions. The clinic is currently using an algorithm to calculate the dose to all its patients that has known inaccuracies when calculating dose in inhomogeneous environments (such as lung next to soft tissue or bone). Dr Garnet conducts a series of measurements and finds that the algorithm overestimates the dose to lung lesions by 2%–5%. His findings agree with those reported in literature. However, he does not have the staff or sufficient accessible machine time to commission a new algorithm. He therefore proposes to start SBRT lung treatments with the agreement that the dose calculated by the planning system will be corrected manually by 5%. He holds a series of training sessions for the planners, the radiation oncologists, and the physicists explaining in detail the shortcomings of the current algorithm and how to go about correcting for them. The first patient is scheduled for treatment two weeks later.

Dr Garnet's position is not an uncommon one. Introduction of new techniques in radiotherapy centres is ongoing. In smaller clinics, this

TABLE 5.9 Scenario 9: Dr Garnet, Introduction of a New Technique

- Dr Garnet is the chief physicist at Sospitas Clinic.
- There is a demand for stereotactic lung radiotherapy at the clinic, both due to the improvement in outcome and the reduction in each patient's treatment time.
- The clinic does not have the resources to purchase, commission and validate a treatment-planning algorithm for correct dose calculation in lung.
- Dr Garnet proposes a manual dose correction for each treatment produced on the current, inaccurate, algorithm.

Dignity Autonomy	Beneficence, Non-Maleficence	Justice	Prudence Precaution	Honesty Transparency
–	(Y)	(Y)	(Y)	(Y)
(N)	(N)	–	(N)	(n)

can be problematic as the costs can be relatively high, staff availability can be limited, and time on the treatment units for testing only means time taken away from patient treatment time. In the case of Sospitas Clinic, the introduction of SBRT treatments, in addition to the improved outcome they offer to their patients, also mean an increase in the number of patients treated as less time will be needed for each patient availing of SABR. For these last two points, Dr Garnet's introduction of the SABR service scores a **(Y)** under both beneficence and justice. His study of the inaccuracies of the current algorithm and his efforts to educate the staff in the same, also give him a **(Y)** under prudence. However, his proposed short-cut solution introduces additional dose inaccuracies as he has no way of evaluating the actual magnitude of error for each patient's calculated dose. This practice adds an **(N)** to beneficence/non-maleficence as well as in prudence. Patients are unlikely to appreciate the nuances of dose calculation and the significance of the potential underdosing even if these are explained to them. Being offered this treatment therefore puts them at risk of harm, and also compromises their consent. In view of the latter, this practice also scores an **(N)** under autonomy. While Dr Garnet's solution is transparent to his colleagues, the clinic's shortcut to the introduction of SBRT and the unlikeliness that the risks will be explained to patients, score both a **(Y)** and an **(n)** under honesty/transparency (Table 5.9).

5.3.10 Scenario 10: Ms Perylene, What Nobody Knows

Ms Perylene has recently joined Medela Clinic as medical physicist. During the first week of her employment, she was asked about her experience with high dose rate (HDR) brachytherapy. Even though she has received some training, her real experience is limited as HDR was not commonly used at the hospital where she was previously employed. However, she felt obliged to say that she has adequate experience so as to not put herself down in front of her new colleagues. Today, the only other physicist with HDR experience is on sick leave, so she is asked to plan the treatment for Mr Greystone, a patient with prostate cancer. She takes time creating the plan for Mr Greystone and when she is satisfied with the dose distribution, she shows the plan to the radiation oncologist who approves it and the treatment goes ahead. The treatment is allowed to proceed without a second dose check as the catheters are already inserted. There is a protocol at Medela Clinic dictating a second check for every plan, so a form for non-conformity is completed and signed by both Ms Perylene and the treating radiation oncologist so the treatment can proceed.

Following the administration of the treatment, Ms Perylene realises that she forgot to include in the plan a 6 mm offset necessary to correct the first source position. She is distressed about the mistake, but feels it is too late to say anything – after all, the treatment has been delivered. She recalculates the dose distribution with the 6 mm offset, and decides that the dose difference is not clinically relevant so there is no need to create a commotion over something insignificant. She makes a note herself to always apply this correction in the future.

This scenario identifies the importance of a second check and the value of honesty and open disclosure in radiotherapy. Unlike many other medications or procedures, there is no antidote or corrective action that can be taken once radiation has been administered. The irreversibility of the effect is one of the reasons radiotherapy accidents make headline news (the spine chilling articles in the *New York Times* are gruesome examples of this (Bogdanich 2010)) and makes the proverb 'measure twice, cut once' particularly relevant. In the case of Ms Perylene, the urgency to treat Mr Greystone because of the inserted catheters superseded the importance of treatment verification, leading to his mistreatment and the violation of the value of non-maleficence. By hiding her mistake, Ms Perylene undermines the appropriate evaluation of the impact of the error and prevents any potential corrective action. The values being violated here are those of honesty/transparency, therefore an **(N)** is added in addition to the **(N)** of non-maleficence. As a physicist, she probably has the experience to assess the clinical impact of a dose discrepancy, but by not consulting with the physician she is acting outside her primary expertise. This behaviour violates the value of prudence adding another **(N)** to Table 5.10.

TABLE 5.10 Ms Perylene, What Nobody Knows

- Ms Perylene has recently been hired as a medical physicist by Medela Clinic.
- When asked, she claimed competence in HDR brachytherapy so as not to put herself down in the eyes of her new colleagues.
- She is now asked to plan an HDR treatment.
- Her lack of competence, and the lack of a second check, lead to the mistreatment of a patient.
- Ms Perylene investigates and decides that the impact of the error is insignificant and therefore does not need to be reported.

Dignity Autonomy	Beneficence, Non-Maleficence	Justice	Prudence Precaution	Honesty Transparency
–	–	–	–	–
(N)	(N)	(N)	(N)	(N)

The behaviour Ms Perylene and the treating radiation oncologist exhibit does not stem from bad intentions. On the contrary, it most likely arises form a desire to 'get the job done'. Professionals acting off protocol are often applauded for their efficiency and agility, when the actions involved do not lead to error. However, prudence dictates that the chances of error must be minimised. Open disclosure of errors honours the autonomy and dignity of the patient, and allows the experience inform future actions. The hesitancy to disclose her errors can be attributed to Ms Perylene's desire to not damage her reputation. This behaviour is not limited to individuals, but is present in whole system. The UK House of Commons Committee in 2011 heard evidence that 'the NHS (National Health Service) does not always admit when things go wrong, nor does it always offer an explanation' (Parliament House of Commons 2011). The Secretary of State's response included the following statement: '...we could strengthen transparency of organisations and increase patient confidence by introducing a 'duty of candour': a new contractual requirement on providers to be open and transparent in admitting mistakes. We agree.' (Department of Health 2011). Transparency does not only increase patient confidence, but also demonstrates respect for the patient. The failure to disclose scores an (N) under dignity/autonomy. In addition, it prevents others from learning from Ms Perylene's mistake, which adds an (N) under justice.

5.4 CONCLUSION

In this chapter we demonstrated how the pragmatic set could be used as a framework for ethical reflection in radiotherapy. The ability to identify, assess and reflect on a situation is of primary importance in the ethics governing radiotherapy as scientific and medical issues can easily overshadow ethical ones. The purpose of this value-based framework, and the analysis of the scenarios based on it, is not to provide answers or directives on how one ought to act. It is rather, an invitation to reflect the common sense morality upon which measures can be built to protect the patient. The process of ethical reflection is discussed in greater detail in Chapter 7. The pragmatic value set does not cover all that is morally important; Chapter 6 re-examines some of the scenarios and expands on this framework. Nonetheless, the five values in their own right add much to explaining and justifying the substantive moral issues encountered in radiotherapy. The addition of prudence and transparency/honesty to the widely applied values of beneficence, non-maleficence, autonomy and justice has proven imperative in radiation protection in radiotherapy. Prudence was necessary in all decisions made in the scenarios when there was uncertainty in the present or future outcome of an action. Honesty and transparency were linked to autonomy and dignity when considering the right of a patient to have all the information he or she needed to make the best decision for managing their health, including how to interpret and manage risk. The role of these two values in establishing a culture of respect amongst the healthcare team members, and by extension, a culture of safety in which the protection of the patient can be optimised, was also highlighted.

Chapter 7 delves further into the process of deliberation, judgment, action and reflection. It should be clear by now that the pragmatic value set is not intended as a procedure for decision making. The circumstances in any healthcare scenario vary greatly from one occasion to another and render it impossible to lay down a set of rules that can direct every action. Rather, the pragmatic set's utility is in helping to systematise our understanding of the values that must be considered.

CHAPTER 6

Extension of the Pragmatic Value Set

6.1 INTRODUCTION

When we proposed our 'pragmatic value set' we based our considerations on the four principles of biomedical ethics proposed by Beauchamp and Childress (1979), but we also noted the similarity of that system with the set of core values identified by ICRP in their report on 'Ethical foundations of the system of radiological protection'. (ICRP 138 2018) We felt that for application in radiological diagnostics and therapy, the Beauchamp and Childress set should be complemented by two additional values, which turned out to be one of the core values and one of the procedural values of ICRP. The complete set of five values (or six, if we list non-maleficence and beneficence separately as Beauchamp and Childress do) was intended to be comprehensive enough to address most ethical questions in medical radiology. We tried to demonstrate this in the preceding chapters.

The cross-cultural approach taken (among others) by ICRP permitted the identification of additional principles/values that might be, perhaps not throughout, but just in specific cases, relevant to radiological protection in medicine. One of these was 'dignity', which the Commission suggested as a core value rather than 'respect for autonomy' (ICRP 138 2018). As briefly discussed in Chapter 2, dignity can on the one hand be considered the more fundamental concept, but on the other a concept wider than 'respect for autonomy' emphasizing non-discrimination, for instance. In addition to the ICRP report, 'transparency' identified two other procedural values as important, 'accountability' and 'inclusiveness' (ICRP 138 2018). All three will be discussed in more detail below.

Proposals have also been made for public health – without explicit reference to 'cross-cultural ethics' – for principles that would complement those of Beauchamp and Childress. Under the title, 'How Many Principles for Public Health Ethics?', Coughlin (2008) discusses what additional values would be needed if Beauchamp and Childress's approach was to be used beyond biomedicine in the area of public health, and suggests two: precaution and solidarity. Schröder-Bäck et al. (2014) reflect on a possible basis for a curriculum of public health ethics in a paper entitled 'Teaching seven principles for public health ethics'. Similar to Coughlin, they do not doubt the usefulness of Beauchamp and Childress' principles, but suggest adding three more: efficiency, proportionality and sustainability. We will briefly look at these five suggestions here:

- Precaution: *Precautionary measures should be taken even if some cause and effect relationships are not fully established scientifically* (Wingspread Conference 1998). When he discusses this principle, Coughlin (2008) writes mainly about sustainability (i.e. the implications of our actions for future generations). We will come back to this question below but would like to point out here that precaution is above all concerned with acting under uncertainty, whether in the future or now, and is therefore a principle separate from sustainability. We have already discussed it under the heading 'prudence' in Chapter 2 and will not go into further detail here.

- Solidarity: *Solidarity or social cohesion … relates to how united, connected, and cooperative a society is* (Coughlin 2008). Although it seems to us that Coughlin relates this principle too closely with a specific school of ethics, communitarianism (see our brief discussion of virtue ethics in Chapter 2), we agree that the interests of society as a whole, the common good, sometimes need to be considered in addition to and separately from those of the individuals immediately affected. We have briefly mentioned the common good in our discussion of justification in radiological protection (see Chapter 2) but will discuss solidarity as an independent principle below.

- Health Maximisation: *The primary end sought is the health of the broader constituency of the public.* Schröder-Bäck et al. (2014) claim that considering beneficence and non-maleficence, and trying to achieve the greatest possible balance between them cannot serve to maximize health in a population, because they focus on the individual. The concern, which in itself we do not dispute, seems to be addressed by giving importance to solidarity, or the common good, as just explained.
- Efficiency: *There is a moral duty to use scarce health resources efficiently.* Schröder-Bäck et al. (2014) discuss the problems of adequately defining efficiency, especially if we do not want to limit ourselves to economic profitability, but quite apart from such considerations it seems to us that there is nothing new in this principle if we have already widened our view from beneficence and non-maleficence for the individual to solidarity and the common good, and if in addition we extend justice for one generation to sustainability for the future (as will be discussed below).
- Proportionality: *In weighing and balancing individual freedom against wider social goods, considerations will be made in a proportionate way.* This is, as pointed out by Schröder-Bäck et al. (2014) themselves, a methodological principle, not really on the same level as the others. We have already discussed in Chapter 2 how one of the challenges of the Beauchamp and Childress approach is to balance the different principles, which all have *prima facie* validity, when they conflict with each other. This is not an easy matter, but it does not seem to be made easier by adding the concept of proportionality, as defined by Schröder-Bäck.

Insight may also be gained by looking at a related field that has some overlap with radiological protection, namely environmental health. We will not go into details here as to the various ethical principles proposed in the literature as relevant, but just note one which did receive some attention from ICRP, even though it is only mentioned in passing, in ICRP 138, 'Ethical foundations of the system of radiological protection':

- Sustainability: *Conduct work in such a way that it meets the needs of both the present and future generations* (Jameton 2010). This is another example of a principle that is not of primary concern in a situation where the focus is on the individual patient. When it comes to the whole health system, however, or to environmental consequences of certain medical technologies, for instance, sustainability is highly important, and cannot be neglected.

Finally, an additional procedural value has been proposed by Zölzer (2016):

- Empathy: *When considering the situation of others, one should take people's emotions and concerns seriously, even if they are not always based on rationality.* This may seem a little out of place for an endeavour that is thought to be based on scientific evidence, even if the importance of value judgements is now recognised in decision making in radiological protection. It is worth noting that empathy has recently received a lot of attention in very practical areas, such as product design, urban planning and other so-called human-centred design processes (Kolko 2014).

Taking all this together, we propose that our 'pragmatic value set' could be, and for some applications should be, extended to a broader set. The five members of the pragmatic value set (or six, if we follow Beauchamp and Childress counting) are: Respect for autonomy, non-maleficence/beneficence, justice, prudence, and honesty. Our suggestions for extension are: Dignity, solidarity, sustainability, accountability, inclusiveness, and empathy. These values can be presented as the four original 'principles of biomedical ethics', plus four correlated principles, plus four procedural principles (Table 6.1 following). But before we discuss the relationships between them, we will have a look at the cross-cultural validity of each.

6.2 A MORE COMPLETE SET OF VALUES FOR RADIATION PROTECTION

6.2.1 Dignity

Dignity is no doubt closely related to Beauchamp and Childress' respect for autonomy. Some authors have even discussed whether the former should not actually replace the latter, but others have criticized this by saying we would replace a relatively well-defined concept by a very vague one. We agree with those who understand dignity as the more fundamental principle and respect for autonomy rather as the derived one, which makes it concrete for certain situations. On the other hand, as the Beauchamp and Childress system is well established, we felt in the context of this discussion it is better to leave the core principles as they are and consider dignity as an additional, correlated one.

Autonomy, namely the ability and right to decide for oneself (especially as a patient), is one aspect of dignity, and non-discrimination on grounds of age, sex health, social conditions, ethnic origin and/or religion is certainly another. Very few people would deny the applicability of this principle to just about any area of human activity. It is expressed in different ways around the world, but the basic idea is virtually ubiquitous – that of a dignity pertaining equally to all humans. In the *Bhagavad Gita*, Krishna says, 'I am the same to all beings ... In a Brahma ... and an outcast, the wise see the same thing'. Similar statements are reported of Buddha and Confucius. In the Bible, the prophet Malachi asks, 'Do we not have one father? Has not one God created us?' The concept is also clearly expressed in the Quranic verse, 'We have conferred dignity on the children of Adam ... and favoured them far above most of Our creation'. And in Bahá'u'lláh's writings we find this: 'Know ye not why We created you all from the same dust? That no one should exalt himself over the other'. (references in Zölzer 2013)

These are just short glimpses from different religious sources, but the broad agreement on the notion that all human beings share the same dignity is also reflected in the 'Declaration Toward a Global Ethic' of the Parliament of World's Religions in 1993. It says that 'every human being without distinction of age, sex, race, skin colour, physical or mental ability, language, religion, political view, or national or social origin possesses an inalienable and untouchable dignity, and everyone, the individual as well as the state, is therefore obliged to honour this dignity and protect it' (Küng and Kuschel 1993).

Moreover, for centuries human dignity has been invoked by secular philosophers. This strand of thought begins with Stoicism, continues through the Renaissance, and leads up to Enlightenment (Kretzmer and Klein 2002). In our time, together with the above-mentioned religious traditions, it has played a very prominent role in the drawing up of the 'Universal Declaration of Human Rights' of 1948 and the 'Universal Declaration of Bioethics and Human Rights' of 2005.

6.2.2 Solidarity

As briefly discussed in Chapter 2, beneficence is mainly concerned with the well being of one particular person – in the medical context – the patient. Beyond that, however, the interest of others affected, or even the general public is certainly also a factor that none of our traditions would disregard. This is what is implied by the principle of solidarity. It has also been referred to as social coherence, or we could say, consideration of the common good.

A particular concern in this context is a situation in which profit and burden are distributed unequally (i.e. the good is provided preferentially to one group of individuals and the harm to another). In this case, we think the cross-cultural approach has indeed something to contribute. Many, if not all, philosophical and religious traditions agree that special attention must be given to the underprivileged. We find a similar way of thinking in Rawls' 'Theory of Justice' (1971), where he states 'social and economic inequalities are to be arranged so that they are to be of the greatest benefit to the least-advantaged members of society'. Rawls is generally considered a deontological philosopher, but in this particular instance we feel his theory reflects 'common morality'.

So, let us again have a look at the primary sources. The *Rigveda* recommends, 'Let the rich satisfy the poor implorer, and bend his eye upon a longer pathway. Riches come now to one, now to another'. The Buddha promises, 'He who pursues wealth in a lawful way, and having done so gives freely of his wealth thus lawfully obtained – by so giving … he begets much merit'. Confucius' counsel is: 'Exemplary people help the needy and do not add to the wealth of the rich'. In the Psalms it is stated that 'Blessed is the one who is considerate of the destitute; the Lord will deliver him when the times are evil'. Of Jesus Christ we read, 'Since you didn't do it for one of the least important of these, you didn't do it for me'. And Muhammad says about the 'doers of good' that they '[would assign] in all that they possessed a due share unto such as might ask [for help] and such as might suffer privation' (Zölzer 2013).

6.2.3 Sustainability

It was mentioned above that precaution is sometimes seen as addressing mainly the problems caused for future generations. While it is true that uncertainties about health effects are usually greater and sometimes of a completely different nature for the future than for the present, the point of the precautionary principle is how to behave under uncertainty in general. The consideration of the well being of future generations, on the other hand, seems to be captured best by the principle of sustainability. More specifically, many authors speak about intergenerational equity. Equity does not mean the same as equality, so we do not necessarily have to treat future generations the same as our own, but we have to treat them as fairly as we can. Sustainability can therefore be considered a corollary to the core principle of justice (for further considerations on the terms justice and fairness, see Chapter 7).

The idea that coming generations have to be taken care of when we make decisions (be it about environmental factors affecting health, or other issues) can claim cross-cultural agreement. Both Hinduism and Buddhism are very much concerned with the idea of *karma*, which sees each thought or action as part of an ever-continuing cycle of cause and effect. In line with this, a Hindu delegation to the World's Parliament of Religion stated, for instance, that 'we must do all that is humanly possible to protect the Earth and her resources for the present as well as future generations', and the Dalai Lama made a similar pronouncement: 'Now that we are aware of the dangerous factors, it is very important that we examine our responsibilities and our commitment to values, and think of the kind of world we are to bequeath to future generations'. The responsibility for those who come after us is expressed somewhat differently in the Torah, where God speaks to Abraham, 'I'm establishing my covenant between me and you, and with your descendants who come after you, generation after generation, as an eternal covenant', and this concept of eternal covenant is equally important for Christians and Muslims. Bahá'u'lláh adds still another component to this by saying, 'All men have been created to carry forward an ever-advancing civilization', which according to a statement of the Bahá'í International Community 'offers hope to a dispirited humanity and the promise that it is truly possible both to meet the needs of present and future generations'. Here we can also mention African customary law, which is aptly summarized by a Nigerian chief as follows: 'I conceive that land belongs to a vast family of whom many are dead, a few are living, and countless hosts are still unborn'. And as an example of recent international documents, we can look at a passage from the report of the United Nations World Commission on Environment and Development of 1987 (Brundtland Commission), which maintains that development must meet 'the needs of the present without compromising the ability of future generations to meet their own needs' (Zölzer 2013).

6.2.4 Accountability

Researchers, regulators, and communicators in radiological protection all carry responsibility towards the stakeholders, even if these are to some degree involved in decision making. Especially when it comes to the negative effects on human health, we will want to hold accountable those who have not done their investigations carefully, who have failed to react properly to the available data, or who have not been forthcoming with information. Anything else would be contrary to non-maleficence and prudence.

Given the emphasis placed by all religions and philosophies of the world on proper behaviour, it would be hard to find any tradition not referring to the actor's responsibility for what he or she did or did not do. To quote a modern representative of Hinduism, Mahatma Gandhi, we have the statement that 'it is wrong and immoral to seek to escape the consequences of one's acts', and Buddha says, 'Don't look at others' wrongs, done or undone. See what you, yourself, have done or not'. Confucius expresses it in much the same way: 'The noble person places demands upon himself, the petty person blames others'. The prophet Jeremiah warns that God will 'give every man according to his ways, according to the fruit of his deeds'. Similarly, the Apostle Paul emphasises responsibility to a higher authority: 'So then each of us will give an account of himself to God'. And an oral tradition of Muhammad contains this statement: 'Each of you is a guardian and is responsible for those whom he is in charge of' (Zölzer 2016).

6.2.5 Inclusiveness

If we ask for the main procedural value behind the much-discussed concept of stakeholder involvement, inclusiveness would seem to be the first choice. Respecting people's autonomy is incompatible with making decisions for them. That would be disregarding their human dignity. Instead, everybody concerned should be somehow included in the decision making – which is the central idea of stakeholder involvement.

It must be admitted that participatory approaches to decision making have historically played a minor role. However, it is possible to point to traditions that consider it highly desirable to solve questions of general interest by way of consultation. Thus it is from one of the oldest sacred scriptures, the *Rigveda*: 'Meet together, speak together, let your minds be of one accord … May your counsel be common, your assembly common, common the mind, and the thoughts of these united' – to one of the newest, the Tablets of Baha'u'llah: 'Take ye counsel together in all matters, inasmuch as consultation is the lamp of guidance which leadeth the way, and is the bestower of understanding'. It is well known that the primitive Christian and Muslim communities provided space for open consultation; an ideal which was soon neglected in both religions' mainstreams, but has been revived, to some extent, at different stages of history in both. The rule of Saint Benedict, for instance, written in the sixth century CE, recommends 'the abbot should consult the whole community in matters of importance, and then come to a decision'. Similarly, in Islam the concept of *shura* (consultation), already mentioned in the Quran, was generally understood to mean that the ruler should turn for advice to others before taking a decision. A relevant statement of Shotoku Taishi, the first Buddhist ruler of Japan, has been quoted above: 'When big things are at stake … many should discuss and clarify the matter together, so the correct way may be found'. Sen, in his 'Identity and Violence' presents evidence that the democratic ideas of classical Greece for centuries found no echo anywhere in Europe, while the form of government in some Asian city–states at the same time can be described as democratic. All this must be considered anecdotal evidence, but it shows that it may be worthwhile looking for participatory approaches in different traditions. At least it demonstrates that the value of inclusiveness is not an invention of modern times and is well compatible with traditions (Zölzer 2016).

6.2.6 Empathy

The term 'empathy' dates from the nineteenth century and as such cannot be expected to be found in much older written and oral traditions. Compassion, loving kindness, and a caring attitude, however, are mentioned everywhere. In the *Bhagavad Gita*, Krishna says: 'Who is incapable of hatred toward any being, who is kind and compassionate, free of selfishness … such a devotee of Mine is My beloved'. Buddha praises 'loving kindness and compassion' as two of the most important attitudes that the believer should cultivate. 'Care for all others' is central to Confucius' teachings. The Talmud contains this statement: 'Loving kindness is greater than laws; and the charities of life are more than all ceremonies'. And in one of the epistles ascribed to the apostle Peter we find this exhortation: 'Be of one mind, sympathetic, loving toward one another, compassionate, humble'. An Islamic oral tradition relates that Muhammad said to his followers: 'You won't be true believers unless you have compassion, and I am not referring to the mercy that one of you would have towards his companion or close friend, but I am referring to mercy or compassion to all'. And an American Indian Proverb recommends, 'Never criticise a man until you've walked a mile in his moccasins' (Zölzer 2016).

6.2.7 Summary

A summary of our proposal for core, correlated and procedural principles is given in the following table. It has to be admitted, of course, that the strict one-to-one associations suggested by this form of presentation are untenable. As indicated, sustainability has to do with beneficence as well, not only with justice, and is also closely related to precaution. Solidarity is a matter as much

TABLE 6.1 Our Suggestion for a Largely Complete Set of Values Which Should Be Taken into Consideration When Addressing Ethical Questions of Radiological Protection

Core Values	Correlated Values	Procedural Values
Respect for autonomy	Dignity	Inclusiveness
Non-maleficence	Precaution	Accountability
Beneficence	Solidarity	Empathy
Justice	Sustainability	Transparency

Note: Based on the Beauchamp and Childress (2013) four values (or principles) of biomedical ethics, correlated with the values from public and environmental health, and adding procedural values mainly as suggested by ICRP. (From ICRP 138 et al., *Ann. ICRP*, 47, 2018.)

of beneficence as of justice, and it should not only be practiced in correlation with beneficence, but also with non-maleficence. Inclusiveness could be discussed as being based on justice instead of human dignity, and vice versa transparency would seem to follow from human dignity almost as much as from justice. The table is thus a useful aide memoir to be taken with a pinch of salt, just showing some essential relationships, as well as (hopefully) lending some structure to the ideas put forward in this chapter (Table 6.1).

From the foregoing, it seems clear that not only the four principles proposed by Beauchamp and Childress, but a number of additional principles, be they corollaries or extensions of the original four, or applications in terms of procedural ethics, are indeed based on values which are shared across cultures. They can be traced back to the religious and philosophical traditions that have provided moral guidance for people around the world over the centuries. That is not to say that secular ethics is wrong and useless, but just that a degree of worldwide consensus already exists and is reflected in those traditions. Whether radiological protection in practice has always and everywhere reflected these values is a different question, but there is a growing awareness of their importance.

In Section 6.3 below we will have a second look at some of the scenarios presented in Chapters 4 and 5. We will assess if taking into consideration one or more of the values discussed here in addition to the 'pragmatic value set' further corroborates (or maybe contradicts) the verdict reached earlier. While the range of examples we can look at is not comprehensive, we will see that in many cases it is not so much the decision for the individual patient for which the additional principles give guidance, but rather aspects of the overall organisation of radiological diagnostics and radiotherapy.

6.3 THE EXTENDED LIST AND SCENARIOS FROM CHAPTERS 4 AND 5

6.3.1 Dignity in Earlier Scenarios

Respect for autonomy, and also dignity have been discussed in all the cases presented in Chapters 4 and 5. The additional aspect of non-discrimination, however, has not been touched upon in those exercises. It may seem at first that it is so self-evident that we do not need to speak about it. Nobody would admit that he or she took a decision differently because the patient was a foreigner, had a skin colour different from most other patients, belonged to a different faith group (visible from her headscarf or his turban), or indicated a sexual orientation different from the majority. In real life, it may well happen that individual doctors allow themselves to be influenced by such factors, but the societal climate is increasingly such that nobody would admit to it. We may well ask, however, whether there is not (in some countries) an 'institutional' discrimination in favour of patients with a certain type of medical insurance, and against patients who are not part of the insurance system at all. This may be further aggravated by financial, reimbursement or remuneration arrangements for hospitals, clinics, practices, professions and their individual members.

In case of a procedure usually not covered by medical insurance, it may simply be the economic situation of the patient that decides. Considering Scenario 8 reproduced here as Table 6.2 (Chapter 4, Section 4.4.1 for

TABLE 6.2 Scenario 8: Dr Salmon, Cardiologist in Private Practice

- Dr Salmon, Interventional Cardiologist. Private rooms with associated imaging facility.
- Explains the radiation (and other) hazards of procedures. Explains radiation risk is unproven.
- Accepts *IHA* and unreferred worried well.
- Procedure on request with consent.
- Fee for consultation, separate charge for imaging.
- Dr Salmon is shareholder in imaging facility and does not advise patients of her financial interest.

Dignity Autonomy	**Beneficence, Non-Maleficence**	**Justice**	**Prudence Precaution**	**Honesty Transparency**
(Y)	**(–)**	**(–)**	**(–)**	**(Y)**
(n)	**(N)**	**(N)**	**(N)**	**(n)**

Source: Reproduced from Table 4.8 in Chapter 4.

full discussion), we have to take into account that Individual Health Assessment (IHA) for symptom-free patients is something not generally affordable. In addition, the extent to which it is practiced is sensitive to cultural differences and subject to great variation throughout the world (Malone et al. 2016).

In practice, IHA in many countries will only be available to better off persons, or those holding special insurance arrangements, often through an employer. Only these will have freedom to access it and 'benefit' from its limited and somewhat questionable diagnostic possibilities. But they will also be exposed to higher probable risk with little prospect of benefit (see Section 7.2.3). As mentioned above, these risks may outweigh the benefits, so it seems that in this case doctors offering IHA could discriminate against the well-to-do and expose them to higher risks.

The opposite might be true when it comes to certain forms of radiotherapy. In some countries at least, not every insurance is willing to cover the costs of proton radiotherapy, even if doctors consider it advantageous in particular cases. Patients then have to pay from their own resources. Here, the system is discriminating against the less well-off as it may do in many other cases of advanced and expensive medicine or health care which is not available to the poor or those with a certain basic form of insurance (in those countries which have a kind of 'two-tier medical system').

Similar considerations apply in Scenario 5 in Chapter 4. Here, the younger members of the population receive higher-dose CT examinations on an old machine, while older patients, generally with better insurance, are selectively directed to a new low dose machine (see Section 4.2.5 for full discussion).

One might think that all this is a matter of justice rather than dignity/non-discrimination. It certainly can be discussed under that heading, but we would like to reiterate that the definitions of justice are diverse, and some do not have any difficulty with a health system that is based on inequality. If some people contribute to society more substantially than others and therefore have a higher income – so some may argue – it is only fair and just that they should have access to more and possibly better health services. The WHO constitution, however, considers '... the highest attainable standard of health as a fundamental right of every human being', which makes the whole question one of non-discrimination above all (WHO 2006).

Radiology and radiotherapy both rely on expensive technologies and highly trained specialist staff with highly varied remuneration systems throughout the world. The levels of service available in different countries, in their public and private systems, provide examples of recognizing, and failing to recognize, values such as dignity, or non-discrimination on an economic basis. Perversely, the more florid and abusive systems of reimbursement can leave both rich and poor at disadvantage. This occurs when the ethical sensitivity is downgraded, and additional healthcare spending is often misdirected with poor overall outcomes (Wennberg et al. 2008; Papanicolas et al. 2018).

6.3.2 Solidarity in Earlier Scenarios

It was noted earlier that solidarity is an extension of beneficence. It suggests that we should take account not only of the good of the individual patient, but also of the common good. From this perspective, some of the cases of radiotherapy discussed above deserve a second look. Scenario 2 in Chapter 5 acquainted us with a patient who is very much focused on having his radiotherapy done in a hospital close-by, although a somewhat more distant hospital is offering an equally effective therapy, which is also less sophisticated and would not further stress an already congested facility at the nearby hospital (see Chapter 5, Section 5.3.2 for full discussion). The doctor decides to not offer the alternative treatment at the closer hospital to the patient and thus in a way disrespects his autonomous decision making. This was discussed above as a reflection of the principle of justice. But again one could perhaps debate the exact meaning of justice. The principle of solidarity, however, should make it clear even to the patient that he should not demand special services and time, where this distracts from the needs of other patients, even if the solution offered to him is a bit less convenient.

Solidarity, in the sense of considering the common good, also clearly plays a role with those cases involving 'quasi medical' non-medical exposures. While the pragmatic values are focused on the individual undergoing a radiological examination, the principle of solidarity points to the larger context in this case. In the case, described in Table 6.3, which involved a suspected drug smuggler (drawing on scenario, Chapter 4, full discussion in Section 4.4.2), solidarity actually seems to work against the conclusion reached before. The rights and needs of the individual quite often clash with the rights and needs of the community/general public, and we do not have a patent solution here. This is not an uncommon occurrence. For the purposes of this book, we are content with pointing

TABLE 6.3 Scenario 9: Ms Auburn, Drug Smuggling Suspect

- Ms Auburn (28) at Airport. Sniffer finds small amount of drugs in hand luggage. Suspicion she may be mule.
- Refuses permission for CT scan. Customs officer insists, and hospital does the scan. She is pregnant and no drugs.
- Issues rescan, around Justification, benefit to individual, confidentiality, consent, and governance arrangements.
- Hospital-based scan and lack of clarity on who justifies.
- Risk of undermining privileges of medical procedures.

Dignity Autonomy	Beneficence, Non-Maleficence	Justice	Prudence Precaution	Honesty Transparency
(y)	(–)	(Y)	(y)	(–)
(N)	(N)	(N)	(N)	(N)

Source: Reproduced from Table 4.9 in Chapter 4.

out the dilemma that can occur. As noted in Chapter 2, it seems to us that balancing different values, in this case the good of the individual versus the common good, may be something that is seen differently in different cultures, perhaps in different countries. It must be addressed in a political process, where all sides are heard, and finally a decision is made as to what guidelines the people on the ground – the customs, the police and the hospital staff in this case – should adhere to. Chapter 7 provides further context for this type of discussion.

6.3.3 Sustainability in Earlier Scenarios

It is well known that the embryo, especially in the first trimester of pregnancy, is particularly radiosensitive, and that exposure to radiation at those early developmental stages can cause congenital malformations, reduced mental capacities, and carcinogenesis. Therefore, although the harm or benefit to subsequent generations is usually not a major consideration for radiologists, they are aware of the special risks in situations involving pregnancy or its immediate possibility. Indeed, radiological decisions must be made with a view to both the health of the mother and the unborn child. These concerns are discussed in Scenarios 6 and 7 of Chapter 4 and Scenarios 3 and 4 of Chapter 5 (see Sections 4.3, 5.3.3 and 5.3.4 for full discussion).

But, perhaps the question of sustainability is also a question to the whole population rather than the individual. How we allocate resources now will affect those living in the future, whether in subsequent generations, or within the same generation just perhaps 15 or 20 years down the road. We may also have to consider missed opportunities: the radiologist might sometimes have to ask, 'Do we do our work the way we do it just out of convenience or could we do it differently and establish new and better practices for future patients?'

An example of sustainability along these lines is offered in Scenario 5 in Chapter 5. This looks at the issues involved in patient participation in a clinical trial involving different radiation techniques (see details in Section 5.3.5). On the surface, at least, it does not make much difference for the individual patient which irradiation technique they are assigned to. But, future patients might benefit from the learning that accrues from the trial and formal evaluation of something new. So, the scale is tipped, as it generally is with *bona fide* research studies, by the principle of sustainability, or taking into account the good of other patients in the future.

6.3.4 Accountability in Earlier Scenarios

Accountability can be seen as a procedural reflection of what we discussed in the two preceding sections: Is the practice as it has been established compatible with the common good and with the good of those who will live in the future? Beyond that, of course, accountability plays a role for the question of whether radiation risks have been properly taken account of, whether exposures and doses have been reduced 'as much as reasonably achievable', whether best practices have been followed, including DRL's (diagnostic reference levels) for dose, etc. If one of these is in question, who carries responsibility? Is it clear to every decision maker what his or her accountability is?

Of our examples in Chapter 4, Scenarios 4 and 10 have a particular place for accountability (see full details in Sections 4.2.4 and 4.4.3). In the case of a young doctor in Scenario 4, who makes a wrong referral in a stressful situation, it is hard to find a point in time where accountability should have been more conscientiously brought into the process. However, the cover-up happening later in the scenario is certainly disrespecting this value. Things went wrong, and people should have faced their responsibility and made the situation more public (at least within the context of the hospital, and the requirements of the law), so that others could learn from it. There may be a component of sustainability here – the right of future generations to receive optimal treatment (Table 6.4).

With Scenario 10, the principle of accountability could and should have been brought in earlier during the process. The head of the interventional unit decides to take patients immediately and thus takes over responsibility for something he or she cannot account for. The RPE

TABLE 6.4 Scenario 10: Failed Equipment

- Black Tulip Hospital Interventional Radiology Suite has a tube failure. Urgent replacement by the company three days later.
- Physicist, Dr Russet, contacted to test if system is safe.
- Dr Russet is commissioning a CT elsewhere, and advises he will be available in four days.
- Dr Cinnamon, Head of Interventional, is reassured by the company engineer, and decides to take patients immediately.
- Dr Russet tests the equipment, it appears a filter is missing and exposure protocols incorrect, giving doses ×2 to ×10 high.
- 35 patients receive the high doses. Dr Carmine decides they shouldn't be told and refers problem to RP committee.

Dignity Autonomy	Beneficence, Non-Maleficence	Justice	Prudence Precaution	Honesty Transparency
(y)	(–)	(–)	(–)	(–)
(N)	(N)	(N)	(N)	(N)

Source: Reproduced from Table 4.10 in Chapter 4.

(radiation protection expert) seems to be out of the loop, because he clearly advised that, while he is away, he will be able to do the necessary work in four days. But, as mentioned in Chapter 4, those ordering and making arrangements to install a replacement tube should have informed the RPE that his services will be needed soon. He might then have been able to reschedule his other responsibilities.

6.3.5 Inclusiveness in Earlier Scenarios

The term paternalistic has been from time to time in the discussion here, particularly when a doctor decides without involving the patient. This is of course a failure of respect for autonomy. However, it is useful also to consider the procedural value of inclusiveness in this context. It requires that such decisions are taken only after honest and transparent information is given to the patient, thereby respecting his or her right to know and consent. But inclusiveness also requires that the attitude of the doctor is one of active involvement of the patient. This can raise questions for both the health system and the individual doctor when it comes to allotting enough time to each patient for inclusive decision making. The staffing arrangements the reimbursement model should encourage this approach.

Of the scenarios discussed in the preceding chapters, several show a lack of inclusiveness on the doctors' side, for instance Scenario 3 in Chapter 4, where the doctor explicitly says that she 'deflects further questions (of the paediatric patient's parents) … (because) full explanation takes too much time' (see full details at Section 4.2.3). This is, as we said, certainly not in line with respect for autonomy and dignity, but if the procedural principle of patient participation in decision making had been considered, it would have perhaps been clearer to the doctor that her approach was unacceptable.

A somewhat ambiguous example of inclusiveness is the case of a pregnant patient in need of radiotherapy discussed in Scenarios 3 and 4 in Chapter 5 (see full details at Sections 5.3.3 and 5.3.4). Here, the patient, who is suffering from a borderline personality disorder, is not consistently included in decision making. In Scenario 3, the doctor in charge realises 'she must ensure that her patient understands fully both the risk she places on herself as well as on her unborn child before signing the consent form'. However, in Scenario 4 the team carrying out the CT scan, not knowing anything about her psychological condition, just acts on her once given consent and is unwilling to sort out problems in a process that includes her.

6.3.6 Empathy in Earlier Scenarios

It could be argued that showing empathy to the patient is hardly something that can be demanded of the doctor, as it seems rather a character trait that one has or has not. But, we are of the opinion that at least certain manifestations of empathy can be trained. This is related to the need for the healthcare professionals involved to take time for communication with the patient. The doctor then also has to pose certain questions to the patient, which helps him or her to understand the patient's mind-set or mental state, and to take decisions after having, as far as possible, put himself or herself in the patient's shoes (Table 6.5).

TABLE 6.5 Scenario 6: Ms Violet Pregnant Patient

- Department policy: Ask patient at reception if she is pregnant.
- Does not have more rigorous approach as this would be 'time consuming' and/or 'inconvenient'.
- Ms Violet has elective pelvic CT and is found to be pregnant.
- She is happy to be pregnant, but very distressed about the radiation, does not trust the advice she receives, and considers seeking a termination.

Dignity Autonomy	Beneficence, Non-Maleficence	Justice	Prudence Precaution	Honesty Transparency
(y)	(y)	(–)	(–)	(y)
(N)	(N)	(–)	(N)	(N)

Source: Reproduced from Table 4.6 in Chapter 4.

A case where empathy as an independent procedural principle could have been helpful is described in Scenario 6 of Chapter 4: a patient undergoes a CT scan, and later turns out to be pregnant. She is actually asked at the reception whether she is pregnant and denies it, but the point is that a more empathetic approach would have avoided this interview about pregnancy 'in a relatively public place' (as we pointed out earlier). Similar considerations apply to various aspects of the case, involving a woman on fertility treatment, in Scenario 7, Chapter 4 (see full details of both scenarios in Sections 4.3.1 and 4.3.2).

All this is not to say that efforts to include empathy are always positive and free of problems. Sometimes, for instance, out of (misguided) empathy with the patient (or with the parents, in case of a child patient), full information is withheld and therefore honesty is compromised. But the need of balancing values that are in conflict, is something we have pointed out from the beginning.

6.4 CONCLUSION

It is clear that by referring back to the scenarios from the preceding chapters, the examples illustrate why and how it may sometimes be useful to go beyond the 'pragmatic value set'. We still think that in most cases of medical radiology and radiotherapy, the original pragmatic set of five values will be sufficient, but would like to emphasize that this is not a one-fits-all solution for every possible ethical dilemma, and the additional values suggested here offer the opportunity of more nuanced individual solutions as well as suggesting social situations that require further exploration and analysis.

CHAPTER 7

Reflections on Uncertainty, Risk and Fairness

7.1 ETHICS, FAIRNESS AND TRUST: THE IDEA OF FAIR RISK GOVERNANCE

7.1.1 Understanding Risk-Inherent Technology, from an Ethics Perspective

Science and technology have dramatically changed our world in the last centuries, albeit in conflicting ways. On the one hand, they have significantly contributed to the improvement of our individual lives, our collective well being and the organisation of our society. On the other hand, they have resulted in various threats to life and well being and provided multiple tools to distort and even destroy our society and habitat. The development and application of modern science and technology in the various 'sectors' of our society (health, food, water, housing, energy, transport, industry...) can be called one of the five evolutions that, in a historical perspective, made up modernity. The other four happened in the 'fields' of politics (the emergence of democracy, the nation state and international politics), economics (the emergence of globalised markets and the financial economy), culture (the emergence of popular culture and modern and postmodern art) and the social (the emergence of new lifestyles and new forms of communication).

Evaluation of how science and technology (might) affect our lives and co-existence in positive and negative ways cannot be done in isolation from the contexts within which they operate, which means they must take account of all aspects of the fields of politics, economy, culture and the social as mentioned above. The reason is that the potentialities and (possible) threats of science and technology affect the way we live but also our considerations on the way we want to live. Conversely, current political, economic, cultural and social dynamics affect the way science and technology develop and the way they are applied now and in the future. The awareness of the need to understand, study and govern science and technology in their broader societal context has inspired and instructed general research and policy approaches such as 'technology assessment' (TA), including Health Technology Assessment (HTA), 'science and technology studies' (STS), 'risk governance', 'sustainability assessment' and 'transdisciplinarity' and more specific idioms such as 'second mode science' (Gibbons 1994), 'the co-production of science and social order' (Jasanoff 2004), 'well-ordered science' (Kitcher 2011) and 'post-normal science' (Funtowicz and Ravetz 2003). Meanwhile, the normative aspects of and similarities and differences between these approaches have become topics of research in themselves (see, among others: Carrier et al. 2014; Jordan and Turnpenny 2015).

Many researchers practicing TA, HTA or STS aim to present an 'objective' sociological picture of the interrelation between science, technology and society. Yet, visions such as those of Kitcher and of Funtowicz and Ravetz are normative-driven from the start. They present critical visions on what science should be and on how technology should be understood and governed taking into account the complexity of modern society and they consequently formulate the ethical consequences of these visions. This chapter is written from a similar critical perspective.

What do we talk about when we talk about ethics? Ethics is concerned with questions of right and wrong. But there are different 'levels' of thinking about these questions. Philosophy identifies 'meta-ethics' as the discipline or perspective that deals with concepts of right and wrong (what is rightness? what is goodness?). Next to that, philosophers speak of 'normative ethics' as the discipline or perspective that considers the points of reference that can be used to evaluate a specific practice or conduct. In that sense, normative ethics refers to 'what ought to be' in absence of 'evidence' that would facilitate straightforward judgement, consensus and consequent action. The missing evidence can refer to knowledge-related uncertainty, due to incomplete or speculative knowledge (including scientific knowledge), or due to the absence of an undisputed law or an 'absolute' (set of) value(s) to guide behaviour or choice. All of these apply to the case of the evaluation of risk-inherent technology in our society today. The idea developed in this chapter is that anyone with a specific interest in relation to a risk-inherent technology becomes a moral agent and has a specific responsibility in dealing with that technology in a 'fair' way.

How to understand risk governance? In the context of this book, a risk is generally a *health risk* arising from specific practices (i.e. the application of radiation in medicine). In medicine, health risk governance may refer to policy practices such as provision of public hospitals with radiological facilities to be deployed in a framework determined by legal and good practice protocols for health benefits to all individual members of the community. Within this, there will be provisions for screening or diagnostic mammography services and regulations guiding the use of specific technologies such as CT scanning.

Risk governance can today be seen as a general approach to research and policy related to risk, as it is developed in the fields of science and technology. The scope of the governance includes risk assessment methods and tools, the theory and practice of risk perception, communication and consequent 'governance' (see, among others: Renn 2008) and the ethics of dealing with technological risk (see, among others: Asveld and Roeser 2009). In this chapter, we take risk governance to be a normative approach to practical dealing with risk that first includes *the act of judging* whether a specific risk is justified as acceptable in a specific situation, and that, as a normative approach, has attention for the ethical aspects that inspire and instruct that act of justification (taking into account values such as autonomy and transparency).* Once a risk is justified (acceptable), risk governance moves to its management (optimisation, dose limitation), its assessment of how the risk will eventually manifest itself in practice, and its re-reassessment when necessary. In view of all this, the idea that will be elaborated in this chapter is that 'fair' risk governance refers to *the fairness of the method of risk governance*, considering the challenge of dealing with the 'missing evidence' in risk assessment as formulated above.

As mentioned in Chapter 1, this book proposes an ethical framework for radiological protection suited to its application in medicine. This takes

---

* It is important to note here that 'justification' is understood in the way it is defined by the ICRP and used in Chapters 2 through 5 of this book. The aim here is to stress that justification should be seen as an act of decision making and that the method used to do this decision making should be subject of normative thinking.

the form of the 'pragmatic value set' developed in Chapters 2 and 3 and extended in Chapter 6. In addition, this chapter will situate the pragmatic set in a perspective of making sense of the many perplexing problems encountered in practice. The idea is that, given that assessment of health risks arising from diagnostics and therapy needs to take into account knowledge-related uncertainties and value judgments, there is a need to consider the ethics of *why* and *how* these uncertainties and value judgements can be taken into account. The following text will hopefully make clear how this should be seen as an ethical perspective in addition to the ethics approach elaborated in the previous chapters.

7.1.2 Justifying Risk: Concepts of Fairness and the Idea of Intellectual Solidarity

Taking account of its various applications, the use of radioactivity probably represents an extreme case of how science and technology can serve both cure and be the source of serious damage and even destruction. While medical applications of radiation save individual lives every day, nuclear weapons have the potential to destroy humanity as a whole. Nuclear energy for electricity production, being the other major application of radioactivity, has benefits as a low-carbon source of electricity, but radioactive waste management and disposal remains a complex technological, social and political challenge and a nuclear accident can have dramatic impacts on the environment and on the physical and psychological health of a population for a long time. The specific case of nuclear technology for electricity production is also an extreme example of how technology assessment can be troubled by the fact that 'benefits and risks' of a technology are essentially incomparable (see Chapter 3). From a philosophical perspective, we could say that, due to the specific character of the nuclear energy risk, its societal justification is troubled by moral pluralism. That is: even if all agreed on the scientific knowledge base for the assessment of the risk, then value-based opinions on its acceptability could still differ. Science may thus inform us about the technical and societal aspects of options, it cannot instruct or clarify the choice to make. The matter becomes even more complex if we consider the fact that, in this case, science can only deliver evidence to a certain extent. Nuclear energy science and engineering have reached a high level of sophistication, but we have to acknowledge that the existence of knowledge-related uncertainties imposes fundamental limits to understanding and forecasting technological, biological and social phenomena related to risk assessment and governance of nuclear energy.

Last but not least, we have to accept that important factors remain to a large degree beyond control. These are human behaviour, nature, time and potential misuse of the technology.

In contrast to nuclear energy technology, the use of radiation in medicine presents a different picture. While the evaluation of the eventual construction of a specific nuclear power plant in a specific location *always* concerns society at large (at a regional, national and global scale) and multiple generations in the future, the evaluation of the use of radiation in a concrete medical diagnostic or therapeutic practice primarily concerns a smaller circle of people (the patient, the medical doctor, the relatives of the patient). Obviously, general policies concerned with the overall (increasing) population radiation burden that come with diagnosis and therapy (as highlighted in Chapter 1) also concern society at large and, to some extent, the future generations. As an example, the question whether mammography campaigns as a health policy practice are justified or not is an ethical question that needs to take into account moral pluralism considering (often incommensurable*) values regardless of the available scientific knowledge. Also, in this case, science may inform us about methodology and technical and societal aspects of options, but it cannot instruct or clarify the choices to be made except in a very narrow basis that excludes many important perspectives.

The resulting room for interpretation complicates risk assessment in both the context of nuclear energy and that of the use of radiation in medicine, and puts a specific responsibility on science and technology assessment as a policy-supportive research practice and on practitioners (such as radiation protection professionals) in concrete cases. In simple terms, that responsibility comes down to acknowledging and taking into account uncertainty and pluralism as described above, and the consequences thereof for research, policy and practice.

Similar to the cases of nuclear energy technology and medical applications of radiation, one may understand that the evaluation of risky practices in general may be influenced by moral pluralism, in the sense that judging whether a practice would eventually be acceptable can usually be done with reference to 'external' values. If we thus consider that an evaluation of the acceptability of a risk-inherent practice in general depends on knowledge-based opinions and values-based opinions, we can then construct a simple

* 'Incommensurable' is a philosophical term, meaning incomparable in the sense of 'not able to be judged by the same standards; having no common standard of measurement'.

picture of four distinct cases as presented in the table below. The table may be oversimplified in the sense that one cannot always 'separate' knowledge from values but it can be used as a meaningful tool to determine key concepts of fairness of risk assessment and governance and to understand differences between risky practices in that respect (Table 7.1).

At this point in the reasoning, it might be good to specify what is meant with 'fairness' and 'intellectual solidarity' in the table, and in this chapter in general. The Oxford dictionary defines fairness as 'Impartial and just treatment or behaviour without favouritism or discrimination'. Interpreted in the context of this chapter, fairness thus means that the evaluation of the acceptability of a risk should be done impartially, not favouring or discriminating people in one way or another.* However, as a prerequisite for fairness, *all concerned* first *need to recognise* the existence of knowledge-related uncertainty and value pluralism. Recognising this comes down to recognising the 'limits' to one's own 'authority' when it comes to judging the acceptability of the risk, and recognising the 'right' of others to also have their say in this judgment. This 'triple' recognition has a symmetric and mutual character and may be referred to as a form of 'intellectual solidarity'.

The meaning of fairness (and its relation to justice) and of intellectual solidarity will be further developed later in the text. For now, the table shows primarily that the risks of bungee jumping, mobile phones, nuclear energy or the use of radiation in medicine are not comparable, as the evaluation of their acceptability depends in different ways on knowledge and values.

The bungee jumper will not ask to see the test procedures of the rope before making a jump. In general, the jumper trusts that these ropes will be ok, but more importantly, he or she makes the decision to jump on a voluntary basis. Even though more than one million people die annually in car accidents globally†, no reasonable person advocates a global car ban. Likewise, with bungee jumping, the key concept of fairness related to taking the risk are precaution, informed consent and fair play. In the case of car driving, precaution not only refers to protection measures such as air

* Note that this includes people potentially affected by the risk (such as citizens or patients) *as well as* people with specific responsibilities related to the risk (scientists, politicians, radiation protection officers, doctors…). In this context, impartiality thus means that no distinction can and should be made 'within' these groups and neither *between* the two 'groups'.

† The World Health Organisation (WHO) Global status report on road safety indicates that worldwide the total number of road traffic deaths remains unacceptably high at 1.24 million per year (World Health Organisation 2015).

TABLE 7.1 Justifying Risk – Mapping the Field

		Value-Based Assessment	
Risk-Inherent Practice Acceptable?		**Dissent** **'Moral Pluralism'**	**Consent** **'Shared Values'**
Knowledge-based assessment	**Uncertainty** (incomplete and speculative knowledge)	Governance by **deliberation** *Examples:* • Nuclear energy • Fossil fuels • Official medical screening programmes, e.g. mammography *Fairness:* • Caring for 'intellectual solidarity' in dealing with incomplete and speculative knowledge and moral pluralism ↓ *Key concepts* • Precaution • Informed consent • Transparency • Confrontation of rationales • Accountability to future generations	Governance by **pacification** *Examples:* • Medical radiation applications (diagnosis and therapy) • Mobile phones • Smoking *Fairness:* • Caring for 'intellectual solidarity' in dealing with incomplete and speculative knowledge ↓ *Key concepts* • Precaution • Informed consent • Transparency • Confrontation of rationales • Accountability to future generations
	Consent (consensus on 'evidence')	Governance by **negotiation** *Examples:* • Fossil fuels *Fairness:* • Caring for 'intellectual solidarity' in dealing with moral pluralism ↓ *Key concepts* • Precaution • Informed consent • Confrontation of rationales • Accountability to next generations	Governance by **'simple' regulation** *Examples:* • Traffic • Bungee jumping *Fairness:* • Caring for 'intellectual solidarity' in our behaviour towards each other ↓ *Key concepts* • Precaution • Informed consent • Fair play

Source: Adapted from Hisschemöller, M. and Hoppe, R., *Knowl. Policy*, 8, 40–60, 1995.

bags but also to the value of driving responsibly. And fair play refers in that case to the idea that one can only *hope* that the other drivers also drive responsibly.

The evaluation of the risk that arises from the use of mobile phones or smoking is what one could call a 'semi-structured' or 'moderately structured' problem (Hisschemöller and Hoppe 1995) that can be handled on the basis of 'pacification'. The reason is that, despite the uncertainties that complicate the assessment of those specific risks[*,†], people agree to accept or allow them in light of a 'higher' shared value. This shared value can be a practical benefit (such as in the case of mobile phones) but it may also be a specific freedom (i.e. the choice 'to hurt yourself' in view of a personal benefit, such as in the case of smoking). With reference to the table, one could say that fairness is thus in the way we care for 'intellectual solidarity' in dealing with incomplete and speculative knowledge, and the key concepts of fairness in this sense are precaution, informed consent, transparency (with respect to what we know and don't know and with respect to how we construct our knowledge) and our joint preparedness to give an account of the rationales we use to defend our positions and interests. Because of the uncertainties that complicate the assessment, protection measures are essentially inspired and supported by the precautionary principle. In the case of mobile phones, this principle translates as the recommendation to use them in a 'moderate way' and the recommendation to limit the use by children. For smoking, it translates as anti-smoking campaigns directed at (potential) smokers (with special attention to young people) and as measures to protect those 'passively involved'. Knowing the addictive character of smoking, additional measures are gradually adopted to 'assist' smokers who want to quit.

In a similar sense, evaluating the risk associated with the use of radiation in medical context can also be called governance by pacification. The value

[*] With regard to mobile phone use, the WHO states that 'The electromagnetic fields produced by mobile phones are classified by the International Agency for Research on Cancer as possibly carcinogenic to humans' (World Health Organisation 2014).

[†] With respect to smoking, of course there is the known relation with lung cancer, but the lack of evidence is in the delayed effect and especially in the fact that there is contingency into play (there is no evidence (yet) for why apparently some individuals are more susceptible than others). In addition, while the WHO now clearly states that tobacco kills up to half of its users (World Health Organisation 2018), we don't see these statistics 'happening' in our near social environment. To put it more provocative, our shared values support the idea that we should protect the non-smokers from the smokers, but also the idea that we still live in a free and democratic society where people have 'the right' to smoke themselves to death. It is true that the addictive character of smoking is influencing 'the freedom of choice', but nowadays addicted smokers can always decide for themselves to seek medical and social assistance in their attempt to quit smoking

of informed consent remains central and may also be applied, where necessary, to the close relations of the patient (family members). But, for example, it is generally agreed that the patient takes the risk of a delayed cancer (from a diagnostic procedure) in light of a 'higher' benefit (information/diagnosis about a health condition that will allow it to be better managed).

In contrast to complex problems that are handled on the basis of 'pacification', justifying or rejecting nuclear energy seems to be an unstructured problem that will always need deliberation. Not only do we need to deliberate the available knowledge and its interpretation, but deliberation will also need to consider the various 'external' values people find relevant to judge this case, and the arguments they construct on the basis of these values. Therefore, the fairness of evaluation relates to 'intellectual solidarity' in dealing with incomplete and speculative knowledge but also in dealing with moral pluralism. The key criteria are then again precaution, informed consent, transparency and (preparedness for) a confrontation of rationales. However, they must now also include a sense for accountability towards those who cannot be involved in the evaluation (the next generations). In comparison with nuclear energy, the evaluation of the risk that comes with the use of fossil fuels is a complex problem that, as it would seem at first sight, can be treated on the basis of 'consent on causality'. The Fifth Assessment Report of the Intergovernmental Panel on Climate change states that [...] *Human influence on the climate system is clear* [...] and that '[...] *Warming of the climate system is unequivocal, and since the 1950s, many of the observed changes are unprecedented over decades to millennia. The atmosphere and ocean have warmed, the amounts of snow and ice have diminished, and sea level has risen* [...]' (Intergovernmental Panel on Climate Change 2014). Despite this evidence of a 'slowly emerging adverse effect', the assessment of whether *concrete* draughts or storms can be attributed to human induced climate change or what the *concrete* effect of specific mitigation or adaptation policies would be remains troubled by knowledge related uncertainty. Therefore, in addition, fossil fuel use is a complex problem that requires 'deliberation', and the key concepts of fairness remain the same as for the evaluation of nuclear energy: precaution, informed consent, transparency, confrontation of rationales and accountability to next generations.

Finally, in the medical context, there are numerous radiological policy concerns; some are, and some may not be justified. For example, it is well known that the population and individual radiation doses from CT are greater in Germany, Luxembourg and Belgium, than in most other European Countries (EC 2015b). These questions, and many other concerns, are essentially ethical questions that need to take into account knowledge related to uncertainty and moral pluralism and that, consequently, require deliberation. As in the case of nuclear energy, the fairness of evaluation relates to 'intellectual solidarity' in dealing with incomplete and speculative knowledge but also in dealing with moral pluralism. Also, here the key criteria are precaution, informed consent, transparency and (the preparedness for a) confrontation of rationales.

Before drawing some conclusions based on the discussion of the table, it may be needed to add another comment here on the reason to speak of 'fairness' instead of justice at this stage of the reasoning. The principles of biomedical ethics proposed by Beauchamp and Childress, the set of core values identified by the ICRP and the pragmatic value set developed in the previous chapter, all present and motivate 'justice' instead of fairness as an important value. Reflections on the difference between justice and fairness are topics of philosophical research and debate since the early history of philosophy. Without wanting to make abstraction of that history or of the richness of the debate, in short, one could say that justice is more 'specific' as it refers to specific moral obligations or duties prescribed in procedural rules or guidelines, law and soft law, and supported by a broader societal consensus, while fairness is an assessment of a more general nature still open to various interpretations. Two important interpretations need to be emphasised:

- In a positive sense, fairness can be something 'extra', or 'additional' to justice: our actions can be morally good, but not a requirement of justice, in the sense that we can do things that are morally good while we are not compelled to do them. An example is giving to charity: from a societal perspective, *it can be called fair* that those who have more would give to those who have less or nothing, but there is no specific reference supported by broad consensus that would urge us to give to charity as a moral obligation (Cooke 2014).
- In a negative sense, fairness can refer to something that would 'require' justice, at least from a specific perspective: a patient can claim that *it is not fair* that her opinion on a specific treatment is not asked for. When this sense of (lack of) fairness becomes part of a general sentiment, it acquires a more imperative character and it could eventually lead to procedural rules, guidelines, law or soft law, related to patient participation in decision making.

Referring to the understanding of fairness in Table 7.1 above, it may be clear that fairness is used in the 'positive' sense here. In other words, fairness in the sense of 'intellectual solidarity' is proposed as 'additional' to justice as a requirement to 'enable' justice and the other values proposed. As an example, one could say that it would be 'fair' for a medical doctor to acknowledge towards the patient the existence of uncertainty troubling the evaluation of a specific risk, but under which conditions should this become a moral obligation? The doctor can always claim that, by relying on expertise, she does *not need* to be 'in intellectually solidarity' with the patient (and tell her about the uncertainty), as this would unnecessarily confuse or scare her (see Scenario 1 and 3 in Chapter 4). At the same time, this attitude can also be seen as technocratic and paternalist, and not compliant with the idea of informed consent.

Further in this chapter, the relation between fairness and the values of the pragmatic value set will be commented, and, in conclusion, fairness will also be related to a specific understanding of justice, being the justice of justification.

7.1.3 Three Reflections Re: Ethics, Fairness and Trust in Relation to Risk Governance

The discussion of the table above allows us now to make three reflections related to ethics, fairness and trust in relation to risk governance. Obviously, these reflections are based on our specific understanding of risk assessment in relation to fairness and are therefore presented as a list of ideas that are as such open to discussion:

A. The assessment of what is an acceptable health risk (for an individual, a collective or society at large) is not only a matter of science; it is a matter of fairness in its meaning proposed above, and of values such as those of the pragmatic set. In addition, from the meaning of fairness as proposed above, one can understand that one would want to see the formal inclusion of 'non-scientific values' in decision making in the interest of justice, supported by rules, guidelines, law or soft law.

A.1. A health risk is not a mathematical formula: it is a potential harm that cannot be completely known and fully controlled but is eventually faced in light of a specific benefit, such as a diagnosis or treatment in the medical context. People will accept a risk they cannot completely know and that they cannot fully control only when they trust that its justification is marked by fairness as specified above. And with the meaning of fairness as proposed above, one can understand that it strongly resonates with the value of precaution, and the possibility of self-determination ('informed consent').

A2. Despite the differences between the cases discussed above, they can all be characterised in relation to one idea with respect to self-determination: The idea that 'connecting' risk and fairness is about finding ground between guaranteeing people *the right to be protected* on the one hand and *the right to be responsible* (in the sense of the right to make responsible choices themselves), on the other hand. The right to be responsible depends heavily on the prime criterion of the right to have information about the risk and the possibility of self-determination based on that information. But one must take into account that, in a society of capable citizens, self-determination with respect to risk-taking can have two opposing meanings: It can translate as *the right to co-decide* in the case of a collective health risk (as in the case of mammography or nuclear energy), but also as *the freedom to hurt yourself* in the case of an individual health risk (as in the case of smoking or bungee jumping).

A3. For any health risk that comes with technological, industrial or medical practices and that has a wider impact on society, *the right to be responsible equals the right to co-decide.* And enabling this right is a corollary of justice.

B. Fairness of the assessment of whether a specific health risk is acceptable (for an individual, a collective or society at large) requires a fair dealing with the knowledge about the risk and with the values relevant in the assessment.

B.1. Fair dealing with the knowledge about the risk (knowledge about natural and technical phenomena, causality, likelihood that something will happen, outcome…) comes down to caring for intellectual solidarity in dealing with the uncertainties in knowledge generation.

B.2. Fair dealing with the values relevant in the assessment comes down to caring for intellectual solidarity, recognising the (often incommensurable) value-based arguments those involved would want to bring into the deliberation.

B.3. Taking the previous considerations together, one could say that the assessment of what is an acceptable risk is 'complex', by virtue of knowledge related uncertainty and/or value pluralism. Fair risk governance thus implies *fair dealing with the complexity* of the knowledge and evaluation problems in the assessment of risk.*

C. Trust in the assessment of what is an acceptable health risk (for an individual, a collective or society at large) should be generated 'by method instead of proof'.

C.1. No scientific medical or political authority can alone determine whether a specific health risk would be acceptable or not. Good science and engineering, open and transparent communication and the 'promises' of a responsible safety and security culture are necessary conditions, but they cannot generate societal trust in themselves. The reason is that there will always be essential factors beyond full control, including: nature, time, and human error.

C.2. The fact that people take specific risks in a voluntary way and often base these risks on limited information may not be used as an argument to impose risks on them that might be characterised as 'comparable' or even less dangerous. This principle holds even in extreme cases. For example:

- The fact that the risk of developing cancer from smoking might be 'higher' than that from low-level radiation may not be used as an excuse to impose a radiation risk on people.
- The fact that a professional (such as a radiation protection officer, a worker in a nuclear power plant, a radiologist or a nurse) may voluntarily accept an accumulated occupational dose of 20 mSv per year may not be used as an argument to justify a citizen's dose of more than 1 mSv per year originating from nuclear energy or medical radiation without asking for his or her informed consent.

C.3. Fair-risk governance is risk governance of which the method of knowledge generation and decision making is trusted as fair by society. When the method is trusted as fair, that risk governance

* A more general understanding of 'the complexity of complex social problems' is presented in (Meskens 2016a) and (Meskens 2017).

has also the potential to be effective, as the decision making will also be trusted as fair BY those who would have preferred another outcome.

7.2 AN ETHICS OF CARE FOR FAIR HEALTH RISK GOVERNANCE

7.2.1 Reflexivity/Intellectual Solidarity as Ethical Attitudes in Face of Complexity

From Section 7.1 above, we can now conclude that dealing fairly with the complexity of risk governance in an area such as medical uses of radiation requires joint preparedness of all concerned to adopt a specific responsible attitude. That responsible attitude is *identical* for all concerned (whether doctors, patients, experts* or – in a larger policy context – scientists, hospital managers, entrepreneurs, regulators, advocacy and civil society representatives, politicians, or citizens) and can be described in a threefold way:

a. The preparedness to recognise the knowledge- and evaluation problem and thus the complexity of risk assessment and governance as described above;

b. (following a) The preparedness to acknowledge the imperative character of that complexity or thus to acknowledge one's own 'authority problem' (in addition to the knowledge and evaluation problem) in making sense of that complexity; that preparedness can be reformulated for each concerned participant as the preparedness to see 'the bigger picture and oneself in it', each with his or her specific interests, hopes, hypotheses, believes and concerns;

c. (following b) The preparedness to recognise the importance of intellectual solidarity and, consequently, to seek rapprochement and engage in deliberation with other concerned participants. This deliberation could concern concrete diagnostic and therapeutic protocols of practices in particular, or more generally medical, scientific, technical, technology assessments, health economics, or policy issues in formal interactions in research, politics and education.

* In the context of this text, 'expert' denotes any person with a special expertise as compared to others involved. This could be a scientist in an advisory role towards a political authority or someone who works for a nuclear regulatory commission, but also a medical doctor in relation to a patient.

The threefold preparedness suggested here can be considered as a 'concession' to the complexity sketched out in Section 7.1. From this, a simple but powerful insight flows (i.e. the idea that if nobody has the full effective authority to make sense of a specific problem and its possible solutions), then participants have only each other as the (equal) points of references in deliberating on the problem. In 'The Ethical Project', the philosopher Philip Kitcher reflects to a similar effect saying: 'there are no ethical experts' and that, therefore, authority can only be the authority of the conversation among the participants (Kitcher 2014). From the perspective of normative ethics, we can now (in a metaphorical way) interpret the idea of responsibility towards complexity as if that complexity puts an 'ethical demand' on all concerned, in the sense of an appeal to adopt a *reflexive attitude* in face of the complexity. That reflexive attitude would not only concern the way each participant rationalizes the problem, but also the way each rationalizes his/her own interests, the interests of others and the general interest in relation to that problem.

This responsible attitude can thus be described as a reflexive attitude in face of complexity, and, as a concession towards that complexity – the attitude can also be called an 'ethical attitude' that, in a way, can also be understood as a 'virtue'. However, responsibility also implies rapprochement among concerned participants, and thus in practice, this ethical attitude needs to be adopted *in interaction*. In addition, specific formal interaction methods are required to make that possible. In particular, in a medical case, that implies that a doctor can never make ethical judgements with respect to specific diagnostic or therapeutic practices by him or herself. These judgements should at least be done in deliberative interaction with the patient and (eventually where relevant) the relatives of the patient.

But one can also imagine the need for a wider deliberation, possibly in appropriate established committees or boards concerned with ethics, safety or good practice. With respect to policy of health care practices (such as decisions to engage to a specific level with expensive imaging or radiotherapy technologies and their safety), it is clear that this deliberation also has a broader societal context, outside of the hospital, or the expert committee circle, and involves citizens, politicians and civil society representatives. The joint preparedness for 'public reflexivity' of all concerned participants would enable a dialogue that, unavoidably, will also have a confrontational character, as each would have to be prepared to give an account of his/her interests, hopes, hypotheses, believes and concerns with respect to the problems involved. As briefly suggested before, that joint preparedness can be described as a form of 'intellectual solidarity', as those concerned would have to be prepared to openly reflect with each other and towards the outside world about the way they not only rationalize the problem, their own interests, and the interests of others. Just as understanding reflexivity as an ethical attitude or virtue, *we can also say the same about intellectual solidarity.* In addition, the latter should and could be 'stimulated' by the former. In other words: a sense of intellectual solidarity *implies* reflexivity as an ethical attitude with respect to one's own position, interests, hopes, hypotheses, beliefs and concerns, and this applies to all those involved.

Finally, it is important to emphasise that intellectual solidarity is not an elitist form of intellectual cooperation. It simply denotes a joint preparedness to accept the complexity of health risk governance in general and of specific complex diagnostic or therapeutic practices in particular,* and the fact that no one has a privileged position to make sense of it all. Intellectual solidarity, as an *ethical commitment*, is therefore the joint preparedness to accept that we have no reference other than each other.

7.2.2 An Ethics of Care, 'Bound in Complexity'

As said before, ethics is about judging on 'what ought to be' in absence of evidence that would facilitate straightforward judgement, consensus and consequent action. However, absence of *evidence* does of course not exclude the possibility of some kind of *normative reference* to assist that judgement. Throughout history, philosophers have tried to formulate specific rationales to defend possible references, and one can distinguish four categories of normative ethical theories in Western philosophy in that sense.† Since their emergence at various moments in history, all theories have been subject to academic critique with respect to their

* Although the cases treated in Chapters 4 and 5 are not explicitly described in terms of their character of complexity (the knowledge problem, dealing with uncertainty because of incomplete and speculative knowledge and the evaluation problem, dealing with moral pluralism), one could easily interpret them from that perspective. A further description of these cases from this perspective is outside the practical scope of this book.

† The focus on 'Western philosophy' has no other meaning than to provide a 'pragmatic' framework for the introduction of the ethics of care perspective developed here. Obviously thought from other philosophical traditions and spiritual perspectives, as discussed in Chapter 2, may be relevant here too. However, an elaboration of the concepts of reflexivity and intellectual solidarity from out of those perspectives is outside the practical scope of this chapter.

TABLE 7.2 Dangers and Problems in Western Philosophy Normative Ethical Theories

Western Philosophy Normative Ethical Theories	Danger/Problem
→ **Theories that seek reference in 'universally applicable principles'** ***(Kantian) deontology, consequentialism (utilitarianism)***	Danger: Risk of overlooking the particular of specific situations
→ **Theories that seek reference in evaluating particular situations** ***'particularism'***	Danger: Risk of self-protective relativism (cultural, social, political)
→ **Theories that seek reference in virtues ('being good')** ***Virtue ethics (Aristotle)***	Problem: Virtues do not always unambiguously translate into concrete action
→ **Theories that seek reference in the care for human relationships** ***Ethics of care***	Problem: Works for close relations with known people; unclear how it could work for distant relations with strangers

attempt to universalise their approach. The theories and their critiques can be summarised as illustrated in Table 7.2.

This simple table can now be used as backdrop for the formulation of a specific ethics of care theory that could guide evaluation and action 'in face of complexity' in the context of health risk governance as characterised previously. Why this focus on an ethics of care theory? The argument is that the essence of the theory and practice of moral judgement and ethical behaviour is to be found in a perspective of ethics of care, and this not only for our personal life, but also and essentially for the organization of our more 'formal' interpersonal relations, such as our professional relations in health risk governance.*

Traditional ethics of care is said to work for 'close relations' such as those in a family and with friends. Also the relation between a medical doctor or nurse and a patient and the relatives of the patient can be called 'close', and this is why ethics of care theories are considered more and more as moral guidance for medical practice (see, among others, Branch 2000; Cates and Lauritzen 2002), although skepticism remains (see, among others, Allmark 1995). However, as emphasized in previous chapters, health

* As an additional thought: The basic idea of an ethics of care approach has actually a strong consequentialist character as such: The idea that we would need to judge our actions based on the (potential) direct or indirect effect on our relations with other people, and this based on the understanding that these relations are essential for our existence, which means that we need to 'care' for them.

risk governance does not only refer to concrete practices but also to policies with a wider impact, such as protocols and regulation for use of CT scans. These have an impact on more 'distant' relationships, making it unclear what 'care' might mean. The idea is that the ethics of care theory formulated here might not face traditional problems, as it does not aim to instruct concrete *practical action* of concerned participants but rather inspire specific modes of reflective and deliberative *interaction* among them.

In short, the characterization of complexity as sketched above enables a formulation of an ethics of care that could also work for our distant relationships with strangers. While the previous section elaborated on the meaning of reflexivity and (a sense of) intellectual solidarity as ethical attitudes or virtues, and on the need to adopt these attitudes or to foster these virtues *because* of complexity, the idea is now that, in addition to that, it is possible to develop an ethical theory on how to deal fairly with complexity based on the simple insight that we are all *bound in* that complexity. The idea that 'we are all in it together' informs the view that we should care for our relations with each other, not only in the sense that we need to be reflexive with respect to how our complex relations 'emerge' and 'work', but also in the sense that we *need* each other to make sense of the complexity of issues such as health risk governance, in general, and of diagnostic and therapeutic practices, in particular. The proposal made here is that the 'fact of complexity' brings along three new characteristics of modern co-existence that can be named 'connectedness', 'vulnerability', and 'sense of engagement'. Their meaning in relation to the complexity of complex social problems in general and of health risk governance in particular can be summarised as follows:

Connectedness: We are connected with each other 'in complexity'. We cannot escape or avoid it. Fair dealing with each other implies fair dealing with the complexity that binds us.

Vulnerability: In complexity, we became intellectually dependent on each other while we face our own and each other's 'authority problem'. We should care for the vulnerability of the ignorant and the confused, but also for that of 'mandated authority' (such as that of the scientific expert, the medical doctor, the politician or the teacher). Last but not least, we should care for the vulnerability of those who cannot be involved in joint reflection and deliberation at all. Obviously, without wanting to make evaluative comparisons between them, these can be identified as the next generations, but also as those among us who are intellectually incapable of joining (children, those who may be unconscious or otherwise incapacitated, or with serious mental disabilities).

(*Sense of*) *Engagement:* As modern human beings, our experiences now extend from the local to the global. As intelligent reflective beings, becoming involved in deliberating issues of general societal concern became a new source of meaning and moral motivation for each one of us. As citizens, we want to enjoy the right to be responsible in the complexity that binds us, although not only in our own interest. The idea is that, for contemporary humans, the will to contribute to making sense of the complexity of our co-existence in general, and of the complexity of issues such as health risk governance, in particular, can be understood as driven by an *intellectual need* and as *a form of 'intellectual' altruism*. The contemporary human becomes frustrated and unhappy if she/he is unable to put that social engagement into practice in one way or another. According to the Buddhist thinker Matthieu Ricard, 'real' altruism is a mental attitude, motivation and intention (Ricard 2015).* However, one can understand that acting upon that attitude, motivation and intention will only have limited and temporal effect if at the same time cultures of paternalism, technocracy and conservatism curtail our possibility to engage in practice.

We can now connect this ethics of care perspective with the idea of reflexivity, and intellectual solidarity as ethical attitudes or virtues, as elaborated above. Connectedness, vulnerability, and a sense of engagement, identified as new characteristics of co-existence, imply the need for intellectual solidarity with each other in the way we make sense of complexity of co-existence and of our relations in that co-existence. This can be represented as having a sense for interaction modes that are 'confronting' or 'enabling' at the same time, as presented in Table 7.3.

* Altruism as a 'mental attitude' is of course not a typical Buddhist perspective. Since the concept was proposed by the French philosopher Auguste Comte, the meaning of altruism and the motivations for altruism as an 'attitude' have been the topic of study in philosophy as well as in (evolutionary) psychology and evolutionary biology. For the latter, see, among others (Wilson 2015).

TABLE 7.3 The Idea of Reflexivity, and Intellectual Solidarity in an Ethics of Care Perspective

Connectedness, Vulnerability and a Sense for Engagement Inspire 'Intellectual Solidarity as a Joint Ethical Commitment', in the Sense of	
Connectedness	The joint preparedness To enable and participate in intellectual confrontation with respect to the rationales we use • To defend our interests, hopes, hypotheses, believes and concerns • To relativize our uncertainties and doubts; to recognize that the practical limitations to participation in deliberation cannot be used to question the principle of participation as such;
Vulnerability	The joint preparedness To acknowledge that we are intellectually dependent on each other; to respect each other's authority problem and the vulnerability of those who cannot participate;
(sense for) Engagement	The joint preparedness To enable and support 'intellectual emancipation' of others with the aim of providing every human being with the possibility of developing 'reflexivity as an intellectual and emotional skill', or thus to develop a (self-)critical sense and to be a (self-)critical participant in society.

7.2.3 Enabling Virtues: Intellectual Solidarity in Decision Making, Science and Education

The advanced formal interaction modes to enable reflexivity and the sense of intellectual solidarity referred to above can be given a name and a practical meaning. Reflexivity and intellectual solidarity as public ethical attitudes or virtues naturally would need to inspire the method we use to generate knowledge about the problems faced and the method we use to negotiate and make decisions related to them. So, the question becomes: In what way could these virtues inspire good diagnostic or therapeutic practice and related research and policy making?

It was noted above that the problem with virtue ethics as a theory of normative reference is that virtues do not always unambiguously translate into specific concrete action. Virtues such as being 'good', 'honest' or 'prudent' obviously need to be considered in a practical context/situation to apprehend their practical meaning. But even then, different virtues come into conflict with each other, or acting from the perspective of one virtue can be complicated because of the existence of conflicting values that must be considered. To give but one example: What is the best approach for a radiologist to take when requested to undertake an examination that is classified as

Individual Health Assessment (IHA). This arises when the patient has no symptoms or signs, and no exposure to (work or life related) risks that suggest the examination might be of medical value (see, for example Scenario 8 in Chapter 4). In these circumstances, there is the wish of the individual to have the examination on one side, and possible radiation harms to the individual as well as more subtle and indirect harm to the health sector on the other (Malone et al. 2016). Should the radiologist do the examination respecting the autonomy of the individual? Or should it be refused based on possible harm to the person involved and/or to the health sector?

From the same perspective, it is true that reflexivity and intellectual solidarity don't unambiguously inspire concrete action of concerned professionals, but they can inspire interaction methods to *enable and enforce* them as virtues in the interest of meaningful dialogue. An example may clarify. In Section 7.2.1 above, it was noted that a sense of intellectual solidarity implies reflexivity as an ethical attitude with respect to one's own position, interests, hopes, hypotheses, believes and concerns, and so on. However, to achieve this attitude in practice requires reflexivity as an intellectual and emotional skill, seeing the bigger picture and your part in it. The important thing is that reflexivity as an intellectual and emotional skill may benefit from solitary reflection, but it cannot be 'taught' simply. Neither can it be 'enforced' in the same way as one can do with transparency in a negotiation or deliberation. For all of us, reflexivity as an intellectual and emotional skill essentially emerges as an 'ethical experience' through interaction with others. That interaction may be informal or stimulated by an organisational culture (such as that of a hospital), but one can also imagine that our formal methods of democracy, science and education could be organised so as to allow and stimulate reflexivity to emerge as an ethical experience through 'experiential learning'. In the interest of keeping this text concise, we will briefly comment on how this can be understood for all three of them.

- An advanced method of negotiation and decision making inspired by the ethical attitudes of reflexivity and intellectual solidarity would be inclusive, deliberative, and democratic. It would see deliberation as *a collective self-critical reflection and learning process among all concerned*, rather than as a competition between conflicting views driven by self-interest. This advanced method could inspire healthcare workers about specific diagnostic and therapeutic cases, as well as hospital managements in connection with the working of departments and committees. In a broader health risk governance context, policies and campaigns or recommendations and regulations about imaging and radiotherapy should be subject of deliberation at national and transnational levels, and should be enriched with opinions from civil society, patient advocacy groups and citizens and with well-considered and (self-) critical scientific/medical advice. This advanced deliberative approach would have the potential to be fair in that it would enforce participants to give account of how they rationalise their interests from both strategic and vulnerable positions. And it would be effective as it would have the potential to generate trust based on its method rather than on promised outcomes. While the utopian picture for health risk governance sketched here implies reform of medical, regulatory and radiation protection politics, intellectual solidarity can also free up traditional approaches for the good of society. In addition, on both local and global levels, public participation and deliberation could be organised around concrete issues and the outcome of that deliberation could be taken seriously.

- As the challenge to science in health risk governance in general and with respect to specific complex diagnostic or therapeutic practices in particular comes down to dealing with knowledge-related uncertainty and value pluralism, reflexivity and intellectual solidarity as ethical attitudes inspire a future for science open to visions from outside the traditional disciplines and ivory towers. In other words: creation of knowledge to advise on policy and practice would need to be generated in a 'transdisciplinary' and 'inclusive' way, as a joint exercise of problem definition and problem solving with input from the natural sciences, the social sciences, the humanities and from citizens and informed civil society, including patients and their advocates.

- Last but not least, there is the need for a new vision on education. Dealing fairly with the complexity of health risk governance needs an education that cares for 'critical-intellectual capacity building'. It would be naïve to think that doctors, patients, experts, hospital managers, scientists, politicians, patient advocates or citizens will immediately adopt the ethical attitudes of reflexivity and intellectual solidarity on request. The preparedness of an individual to be reflexive about his/her own position and related interests, hopes, hypotheses, beliefs and concerns is a moral responsibility, but it relies on the intellectual and emotional capacity of the individual to do so. Insight into the complexity of health risk governance in general, into

> diagnostic or therapeutic practices, and an understanding of the ethical consequences thereof for decision making, must be stimulated and fostered in basic and higher education. Education should become pluralist, critical and reflexive in itself. Young people should be given the possibility of developing a (self-)critical mind and a sense for ethics in general and for intellectual solidarity, in particular.

An ethics of care perspective on our modern coexistence 'bound in complexity' provides a powerful reference to defend the value of (and the need for) these advanced interaction methods. It is important to recognise the meaningful relations between an advanced approach to education, research and decision making presented above. Together, they not only enable and stimulate reflexivity and intellectual solidarity based on their discursive potential, but also provide the possibility to foster and generate trust through working with them. That trust considered here is not the trust that the outcome of deliberation will be the 'correct one', but that its method has the potential to be judged as fair by everyone in consensus, given the complexity of the problem.

We have now come to the point in the reasoning where, as suggested in Section 7.1, the ethics perspective related to 'how we make sense of things' can give values proposed as relevant for the use of radiation in medicine their meaning in the interest of practice. The idea is that, in order for all concerned (including practitioners, patients and experts) *to become sensitive* to the values of dignity/autonomy, non-maleficence/beneficence, justice, prudence/precaution and honesty/transparency as a prerequisite to putting them in practice, they need to adopt the ethical attitude of reflexivity related to their own position and related interests, hopes, hypotheses, beliefs and concerns in the first place. In other words: one cannot see the meaning and relevance of values such as dignity, justice and prudence if one is unable to 'see the bigger picture' of the situation and 'oneself in it'. However, as said before, the *ability* to adopt reflexivity as an ethical attitude or virtue depends on whether one is able to develop it as an intellectual and emotional skill. Therefore, considering the perspective on ethics presented in the previous chapters, one can understand that the pragmatic value set would need to be 'extended' to implicitly include reflexivity as an ethical attitude or virtue, and intellectual solidarity as an ethical commitment. This is visualised in the scheme below, and the following section will elaborate on what it implies for radiological risk governance in general and the system of radiological protection in particular (Figure 7.1).

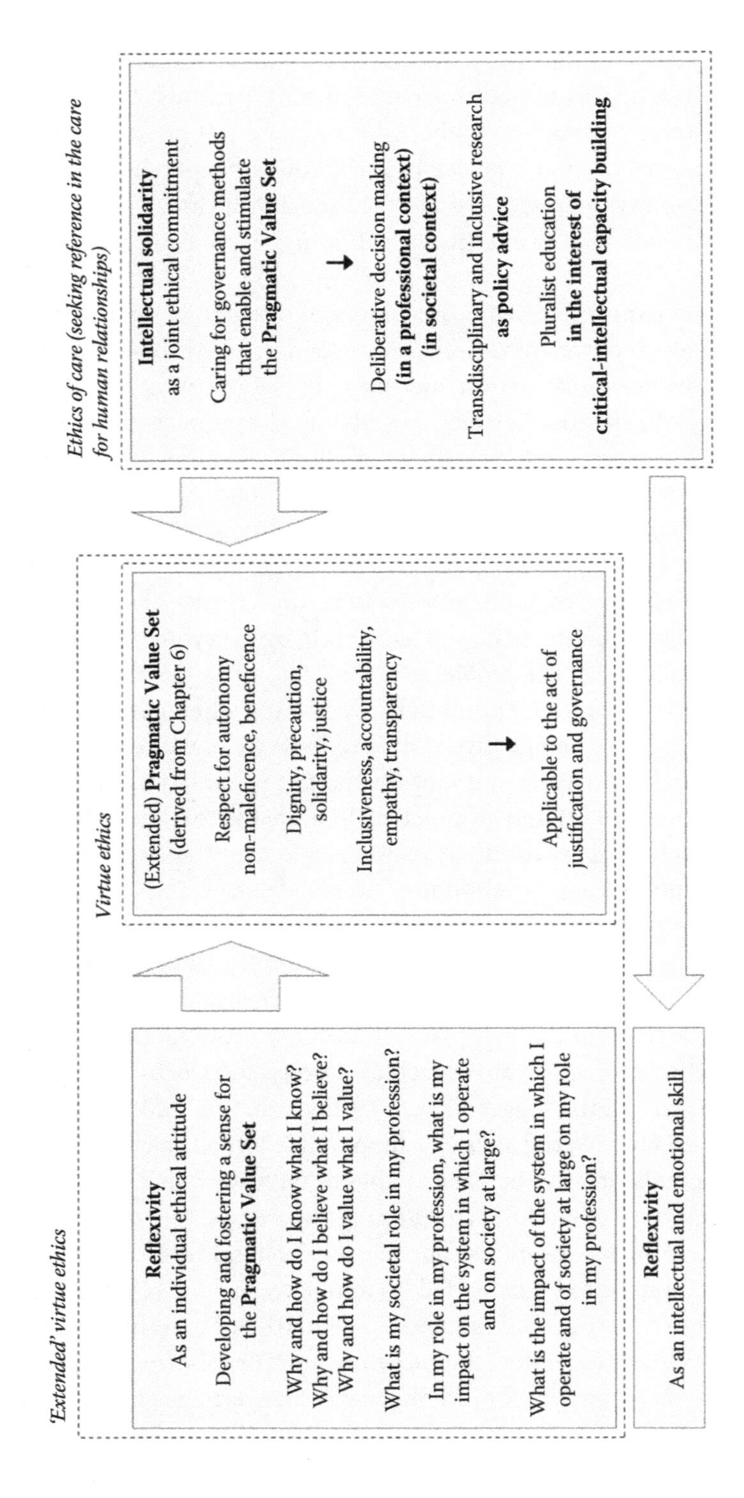

FIGURE 7.1 An Ethical Framework for a Fair Dealing with the Complexity of Justification and Use of Radiation in Medicine.

7.3 CONSEQUENCES FOR RADIOLOGICAL RISK GOVERNANCE

In light of the previous reflections, we can characterise health risk governance in general, and diagnostic and therapeutic practices in particular as complex problems that require their complexity to be dealt with fairly. As a conclusion to this chapter, we may now formulate a set of considerations on how this applies to radiological risk governance.

7.3.1 The Importance of Considering Different Neutral Application Contexts

In Section 7.1, we indicated a difference between the medical and nuclear energy applications context in terms of fairness with respect to dealing with knowledge-related uncertainty (due to incomplete and speculative knowledge) and value pluralism. In both cases, one could call the problem a *complex* problem because its complexity follows the characteristics proposed above. We can also see these are complex *social* problems because they concern the whole range of relevant participants in their application contexts. One may want to 'compare' both application contexts in terms of their complexity, but the essential message here is that a comparison between both technology applications is meaningless in terms of fairness of justification. While ethics for the medical radiation and nuclear energy applications are based on the same or similar principles, the reasons for the values may, on the surface at least, be different. In addition, they may mean different things when it comes to practical application (e.g. when ensuring participation and informed consent of potentially affected persons). This has many implications, for example, in communication: One can observe that nuclear energy advocacy groups like to refer to the advantages of the use of radiation in medicine, in the hope that the public understanding of the humane and valuable contributions of medicine will reflect well on 'their own' context of application.

From a different perspective, one can say that the assessment of the justice of justification of a radiological risk is meaningless *in absence* of a context of application. Consequently, in terms of applied ethics, it is rather meaningless to speak of 'radiological protection', 'radiological risk management' or 'radiological risk governance' without specifying the context of the use of radiation. Even more, it becomes meaningless to speak of 'risk management' or 'risk governance' as such, as, within these distinct application contexts, the radiological risk becomes a joint concern if and only if those involved jointly agree to consider the eventual use of radiation in the context of the proposed 'higher good' (the use of radiation in health care, the eventual use of nuclear energy for electricity production). In other words: only within a neutral governance context (health care, energy governance),* the principles of radiological protection (justification, optimisation and dose limitation) and the values proposed in the extended pragmatic value set receive an ethical meaning that has the possibility to inspire and instruct (as well the required method of deliberation) as to the practical meaning of the values for anyone involved.

7.3.2 Enabling Values in Radiological Risk Governance

Up till now, reflections on ethics in relation to radiological protection, such as those done in ICRP Publication 138 'Ethical Foundations of the System of Radiological Protection' (ICRP 138 2018), have largely focused on virtue ethics. They logically and reasonably follow from the question of what it would imply to be 'responsible' or 'good' as a scientist, manager, policy advisor, medical doctor or regulator. The considerations made in this chapter imply that ethical thinking in relation to radiological risk governance and the radiation protection system requires broader reflection than traditional general or medical ethics alone. It should include ethical reflection on the potentialities and hindrances of the cultures in which those with specific responsibilities are formed and operate. As an example, in Chapter 1, attention was given to the problem of paternalism of the medical professions, arguing that it 'no longer provides an acceptable approach to service delivery and interpersonal behaviour within the services'. Given that cultures are 'self-confirmative' in the way they maintain and protect their own 'comfort zone', one can say that 'paternalism is maintained in a culture of paternalism'. However, cultures such as paternalism are created in themselves through a variety of interlinked processes and practices, and there are reasons to believe that significant 'roots' of this creation process are to be found and confirmed in current approaches to education and research. Looking at how science, education, politics and the market function today, one could wonder how the values identified as relevant for radiological protection in medicine in Chapters 2, 3, and 6, can 'work' in a world ruled by the doctrine of scientific truth and

* The governance contexts are said to be 'neutral' because what is supposed to be 'governed' is neutral (nobody is in favour or against 'health' or 'energy'). But 'within' that governance context, obviously different visions exist on how health or energy should be taken care of.

the strategies of political positionism and economic profit.* It seems as if those values are always in tension with the way science, education, politics and the market function.

Fair radiological risk governance also requires advanced formal governance methods for science, education and political decision making in order to 'enable' these values to work to their full potential. An ethics of care for health risk governance does not only support these advanced governance methods but also gives new meanings to the ethical values underpinning the system of radiological protection. For every professional concerned with radiological protection, whether scientist, engineer, medical doctor, manager, policy advisor or advocacy group, those values receive an enriched 'interactive' ethical meaning when understood as grounded in a care for human relationships 'bound in complexity'. The reason is that acting according to these values always 'starts' with a motivation for rapprochement towards other concerned participants.

7.3.3 The Justice of Justification as a Central Concern

An ethics of care for health risk governance 'bound in complexity' supports the values for radiological protection in medicine proposed in Chapters 2 and 6, as it provides a powerful reference to defend their relevance against cultures of paternalism, the doctrine of scientific truth, the strategies of political positionism and economic profit. However, the ethics of care perspective also supports the idea that precaution and informed consent should not be 'balanced' as trade-offs. Precautionary measures would need to be agreed upon with the involvement of all concerned participants. Therefore, the idea of fair health risk governance integrated in the broader ethical vision of what it implies to deal fairly with the complexity of that risk governance, supports the argument that the justice of justification, including in medicine, ensured by the possibility of self-determination of the potentially affected person or persons (ensuring their 'right to be responsible') should be the central concern of risk governance and of related systems of protection.

In the complex cases of use of radiation (either in health care or energy governance context), a risk cannot be justified through one-directional

* See an extended critique in (Meskens 2016b and 2017) on how the actual governing modes of representative democracy, international politics, science and the market, being modes inherited from modernity as an emancipation process, are no longer able to 'grasp' the complexity of our current complex social problems.

'convincing explanation', but only through mutual agreement among those concerned. An acceptable risk coming with use of radiation is a risk that is eventually justified relying on the formal possibility of deliberation among all informed participants. Obviously, that mutual agreement, an outcome of a justification exercise, can be to either reject or to accept the use of that radiation (see, for example, Scenario 3 in Chapter 4). Therefore, intellectual solidarity as an ethical commitment among all concerned should 'start' with the joint preparedness to see justification as a mutual agreement 'in face of complexity'.

Seen from a different perspective, ethics in relation to the radiological protection system is also about considering and recognising the limits of the radiological protection system when it comes to providing a rationale for justification of a radiation risk. In other words, we cannot question the ethical dimensions of the radiological protection system without also questioning the ethical dimensions of the 'bigger' systems in which the radiological protection system operates and on which it depends. Given that the radiological protection system, in its concern for providing guidance for decision making, relies on science but essentially wants to take into account human and societal values, the bigger systems that need to be questioned in terms of their ethics are those of knowledge production (research, advice), decision making, and also education. For risks that manifest in concrete diagnostic or therapeutic practices, that 'system' is the possibility of deliberative dialogue between the patient, the doctor, the nurse, the radiation control and protection service of the hospital, other hospital agencies, as well as regulatory and professional bodies. For risks that manifest in an occupational context, the system of decision making is the radiation control and protection service and the management system of the organisation, other hospital agencies, trade unions and professional bodies. For risks that manifest on a societal level, that system of decision making is the system of democracy, including the input from citizens, civil society, trade unions, professional bodies, advocacy groups, and of scientific/ethical advice.

This brings us to the need to raise awareness for the possibilities and limits of radiological protection in this sense. In light of what has already been discussed, we can state that fostering a responsible radiological protection culture is a necessary but insufficient condition for the societal justification of the use of radiation in health care or in energy governance. Still many scientists and policy makers claim that a health risk is justified when a responsible regime of protection is put in place. Based on the ethical considerations above, we can conclude that it is the other way around: responsible protection needs to be put in place once all those involved *would eventually* have jointly justified the use of radiation in the envisaged application context.

Finally, taking into account the focus of this book and the above considerations, one can conclude that the radiological protection system cannot and should not be stretched to provide the full rationale for justification of a health risk arising from practices in medicine, but it can and should refer to critical considerations on how our formal methods of knowledge generation and decision making should foster autonomy and involvement of the potentially affected patients and families, and promote vigilance and fairness in justifying radiation risks.

In its recommendations, the ICRP could highlight the importance of the advanced methods for science, politics and education presented above, as a way to ensure fairness in justifying radiation risks, taking into account the different application contexts. In addition, given the central role of science in radiological protection, the ICRP could actively promote a more 'responsible' conception of science as being transdisciplinary and inclusive, not only to advise medical decision makers in assessing justification, but also to support radiation protection policies in both occupational and post-accident conditions. That science would in principle be able to inform policy in a more reflexive and thus deliberate way while it would at the same time be more resilient itself against strategic interpretation of its produced knowledge and hypotheses from out of paternalist cultures, politics, civil society and the market.

Afterword

IN THIS BOOK, WE surveyed the existing ethical frameworks for radiological protection and for medical practice. The principles of ICRP (justification, optimisation and dose limitation) were placed in the context of those proposed for medical ethics by Beauchamp and Childress (autonomy, beneficence, non-maleficence and justice), and we proposed a pragmatic value set specific to medical practice when radiation is involved. The values in the pragmatic set were presented (Chapter 2), applied to healthcare (Chapters 3 through 5), and we explored possible additions to the set (Chapter 6). Chapter 7 examined the broader application of these values in society, emphasizing the need for ethical reflection and judgment where actions must be taken in the absence of complete knowledge of the outcome and of a consensus of visions on what should be done. We have not attempted to proclaim the pragmatic value set as ultimately correct or superior to other frameworks. We have, however, tried to assert its usefulness as a common ethical framework which is necessary to underwrite safe and socially acceptable medical practice, taking due account of the patients' concerns.

The introduction of radiation to medicine in the twentieth century led to advances in healthcare that have benefited millions of lives. A non-invasive assessment of an individual's health has allowed the early detection of conditions that if unnoticed can lead to the detriment for the individual. It has, for practical purposes, made exploratory surgery, once common, a thing of the past. Treatment of disease with radiation offers an option for cure where surgery and drug therapy are either inadequate, ineffective, or cannot be utilised. The changing context in which radiation is used creates new challenges for workers, researchers and the broader healthcare community, who are now confronted with moral questions and ethical dilemmas that were previously confined to energy production and other, even more worrying, uses of radiation.

In medicine, there is a long tradition of professional guidance on the moral dilemmas of the day. At a general level, the declaration of Geneva, adopted by the World Medical Association in 1948, and reissued in revised form in 2017, provides a solid basis for codes of medical ethics. Its fundamental aim includes protection of the dignity and rights of patients and places an onus on the healthcare professional to go above and beyond avoiding harm. Ethically sound medicine must lean on something more than scientific knowledge. It must be rooted in an explicit social contract between patients and healthcare professionals.

A profession exists on the basis of a special body of knowledge that, ideally, is acknowledged, trusted and respected by the public. Patients trust that experts are concerned for the welfare of the public and not just for the respect of their peers. But, professionals and professional bodies do not always treat the expectations of the public/patients with the priority they deserve. This topic generally receives less attention than medical, scientific and political concerns until a crisis involving a public investigation, a court case, or an investigative journalist's report places it in the limelight.

The notion of 'Professionalism' has been added to the medical sciences lexicon and qualities such as compassion, integrity, fidelity, etc. are proposed as its building blocks. The term has gained traction within the medical community. The qualities required of a professional will inevitably vary from one discipline to another. For example, while compassion may be an absolute imperative for the radiation therapist, it may be argued that the same quality is less critical for the radiologist. A value-based framework, as opposed to a quality-based one, may be more successful in unifying the healthcare professionals, building trust with patients, and ultimately protecting them.

The multiplicity of definitions encountered in quality-based professionalism systems inevitably creates a problem. For example, the Irish Medical Council Guide to Professional Conduct and Ethics for Registered Medical Practitioners speaks of the three pillars of professionalism: partnership, practice and performance. But, what is 'partnership' if not respect for autonomy and transparency/honesty between two parties? What is 'practice' if not justice and prudence? What is 'performance' if not aspects of non-maleficence, beneficence and prudence? Similarly, in the UK, the National Health Service lists its values as 'Working together for patients', 'Respect and dignity', 'Commitment to quality of care', 'Compassion', 'Improving lives', and 'Everyone counts'. Although Autonomy, Beneficence and Justice are not mentioned, it is clear that the NHS values are extrapolations of these. These values are also not named per se in the Code of Ethics of the American Medical Association. There is confusion about what we call a virtue, a value, a principle, or things that are simply 'of value' (such as wealth, family, health). The need for a common framework translates quickly to the need for a common language. By adopting a common framework, we can restart the conversation using the same language thus unifying and clarifying the ethical position of healthcare workers in fields utilizing radiation.

Codes of ethics are important in the (often unwritten) social contract between a profession and the public/patient and are intended as a solution to the issue of Babelism in professional ethics. They may summarise the principles, values and qualities that ought to characterize a profession and/or an institution and can set the bar for behaviour that the public has the right to expect. They are often an amalgamation of ethical, legal and managerial guidelines on how professionals ought to act. Codes of ethics, however, do not intrinsically prevent anyone from acting in any certain way. When adopted by organisations, these codes are useful in educating professionals regarding the expectations of their colleagues and the public, and can easily be adopted as a checklist when evaluating a professional's behaviour. A framework based on universally accepted moral values, however, should reach beyond the practical application of a set of rules or guidelines. The ethical conundrums following on the uncertainty in the outcome of medical uses of radiation are manifold. The pragmatic value set can act as a compass and provide a behaviour-defining framework when navigating the multitude of options that arise.

The AAPM's Task Group 109, in 2015, conducted an unpublished survey of 969 respondents (including physicists in clinical practices, research, regulatory bodies and industry). Of these, 49% reported that they had never encountered an ethical dilemma in their workplace. Furthermore, 31.5% reported that they would rely on their personal moral compass to guide their decision if they were to encounter an ethical dilemma. Two main concerns are raised by these responses. First, ethical dilemmas go unnoticed. In practice, areas like radiotherapy and radiology attract numerous very ordinary ethical dilemmas. But they are unnoticed as ways of dealing with them have been absorbed into the culture of the professions and are overshadowed by the concurrent scientific or medical quandaries. This is not uncommon in professions.

Second, there is an over-reliance on the 'personal moral compass' or 'personal ethics' when making choices that directly or indirectly affect patients. This is problematic as moral perception can be affected by many influences, including the prevailing culture, personality traits, personal desires and politico-economic views to name but a few. In the extreme, this approach would allow the moral compass of the cartoon character Homer Simpson to prevail, and surely a professional could not justify this to the public. Such an approach must lead to a corrosion of trust between patients and healthcare professionals. The pragmatic set provides a clear, if limited, set of values, that can be the starting point of a common language shared by radiation protection, medicine, patients and the public. It is limited because, as seen in Chapters 6 and 7, the complexity of the use of radiation in medicine, is such that there is no plausible framework that can produce determined solutions in a full spectrum of cases.

Developing an ethical sense for health care and patient protection is not only a matter for solitary reflection. It must also be grounded in experiential learning, which requires dialogue among all concerned, taking into account everyone's interests, beliefs, hopes and concerns. For health care professionals, this must be a continuous process starting in education and extending through the whole professional career. We have tried to make it clear (without emphasizing it as such) that the values proposed do not concern health care professionals only, but have something to say to everyone holding responsibility towards others in complex matters that can have negative outcomes. The pragmatic value set will, hopefully, inspire a dialogue among all who directly or indirectly contribute to radiation protection of patients.

References

AAPM. 1998. Fraass, B. Doppke, K. Margie Hunt, M. et al. 1998. AAPM radiation therapy committee task group 53: Quality assurance for clinical radiotherapy treatment planning. *Med Phys* 25:1773–1829. doi:10.1118/1.598373

AAPM. 2009. Klein, E. Hanley, J. John Bayouth, J. et al. 2009. AAPM task group 142 report: Quality assurance of medical accelerators. *Med Phys* 36:4197–4212. doi:10.1118/1.3190392

AAPM. 2011. Dieterich, S. Cavedon, C. Chuang, C.F. et al. 2011. Report of AAPM task group 135: Quality assurance for robotic radiosurgery. *Med Phys* 38:2914–2936. doi:10.1118/1.3579139

Abdel-Wahab, M. Fidarova, E. Polo, A. 2017. Global access to radiotherapy in Low- and Middle-income countries. *Clin Oncol (R Coll Radiol)* 29(2):99–104.

ACR. 2013. American College of Radiology Appropriateness criteria. Available at: http://www.acr.org/Quality-Safety/Appropriateness-Criteria (Accessed January 24, 2013).

ACR. 2018. ACR–SPR practice parameter for imaging pregnant or potentially pregnant adolescents and women with ionizing radiation. Available at: https://www.acr.org/-/media/ACR/Files/Practice-Parameters/pregnant-pts.pdf (Accessed August 12, 2018).

Aksoy, I. Elmali, A. 2002. Four principles of bioethics as found in Islamic tradition. *Med Law J* 21:211–224.

Allmark, P. 1995. Can there be an ethics of care? *J Med Ethics* 21:19–24. doi:10.1136/jme.21.1.19

Amis, E.S. Butler, P.F. Applegate, K.E. et al. 2007. American College of Radiology white paper on radiation dose in medicine. *J Am Coll Radiol* 4:272–284.

Appiah, K.A. 2006. *Cosmopolitanism: Ethics in a World of Strangers*. New York: W.W. Norton.

Aristotle. 350 BCE. *Nicomachean Ethics* (transl. 1954 by W.D. Ross). London, UK: Oxford University Press.

Arslanoğlu, A. Bilgin, S. Kubal, Z. et al. 2007. Doctors' and intern doctors' knowledge about patients' ionizing radiation exposure doses during common radiological examinations. *Diagn Interv Radiol* 13:53–55.

Asveld, L. Roeser, S. eds. 2009. *The Ethics of Technological Risk*. London, UK: Earthscan.

Beauchamp, T.L. Childress, J.F. 1979. *Principles of Biomedical Ethics*. Oxford, UK: Oxford University Press.

Beauchamp, T.L. Childress, J.F. 1994. *Principles of Biomedical Ethics.* Oxford, UK: Oxford University Press.

Beauchamp, T.L. Childress, J.F. 2013. *Principles of Biomedical Ethics.* Oxford, UK: Oxford University Press.

BEIR VII. 2006. Biological effects of ionizing radiation (BEIR). *Health Risks from Exposure to Low Levels of Ionizing Radiation: BEIR VII Phase 2.* Washington, DC: National Academy of Sciences Press. Available at: http://www.nap.edu/catalog.php?record_id511340 (Accessed February 16, 2010).

Bentham, J. 1776. *A Fragment on Government.* London, UK: T. Payne.

Bentzen, S.M. Constine, L.S. Deasy, J.O. et al. 2010. Quantitative analyses of normal tissue effects in the clinic (QUANTEC): An introduction to the scientific issues. *Int J Radiat Oncol Biol Phys* 76(3 Suppl):S3–S9.

Bhatia, S. Robison, L.L. Oberlin, O. et al. 1996. Breast cancer and other second neoplasms after childhood Hodgkin's disease. *N Engl J Med* 334:745–751. doi:10.1056/NEJM199603213341201

Bogdanich, W. 2010. Radiation offers new cures, and ways to do harm. *The New York Times.* Available at: https://www.nytimes.com/2010/01/24/health/24radiation.html (Accessed June 1, 2018).

Bok, S. 1995. *Common Values.* Columbia, MO: University of Missouri Press.

Bonn Call. 2016. International Atomic Energy Organisation (IAEA) and World Health Organisation (WHO). *Bonn Call for Action: 10 Actions to Improve Radiation Protection in Medicine in the Next Decade.* Vienna, Austria: International Atomic Energy Agency. Available at: http://www.who.int/ionizing_radiation/medical_exposure/bonncallforaction2014.pdf?ua=1 (Accessed April 19, 2016).

Branch, W.T. Jr. 2000. The ethics of caring and medical education. *Acad Med* 75(2):127.

Brenner, D.J. Hall, E.J. 2007. Current concepts—computed tomography—an increasing source of radiation exposure. *N Engl J Med* 357:2277–2284.

Bryman, A. 2012. Ethnography and participant observation. In: *Social Research Methods.* Chapter 19, pp. 430–467. Oxford, UK: Oxford University Press. On Line resource centre Available at: https://SocialResearchMethods%20(1).pdf (Accessed June 6, 2018).

DOHC (Department of Health and Children). 2006. *Report of the Tribunal of Inquiry into the Infection with HIV and Hepatitis C of Persons with Haemophilia and Related Matters.* Dublin: The Stationery Office. Available at: http://www.dohc.ie/publications/lindsay.html (Accessed May 1, 2007).

Campanella, F. Rossi, L. Giroletti, E. et al. 2017. Are physicians aware enough of patient radiation protection? Results from a survey among physicians of Pavia District - Italy. *BMC Health Serv Res* 17:406. doi:10.1186/s12913-017-2358-1

Carozza, P.G. 2003. From conquest to constitutions: Retrieving a Latin American tradition of the idea of human rights. *Human Rights Quart* 25:281–313.

Carrier, M. Howard, D. Kourany, J. eds. 2014. *The Challenge of the Social and the Pressure of Practice: Science and Values Revisited.* Pittsburgh, PA: University of Pittsburgh Press.

Carver, J.R. Shapiro, C.L. Ng, A. 2007. ASCO Cancer Survivorship Expert Panel. American Society of Clinical Oncology clinical evidence review on the ongoing care of adult cancer survivors: Cardiac and pulmonary late effects. *J Clin Oncol* 25(25):3991–4008.

Cates, D.F. Lauritzen, P. eds. 2002. *Medicine and the Ethics of Care*. Washington, DC: Georgetown University Press.

Chetty, I.J. Martel, M.K. Jaffray, D.A. et al. 2015. Technology for innovation in radiation oncology. *Int J Radiat Oncol Biol Phys* 93(3):485–492.

Chomsky, N. 2008. Anti-democratic nature of US capitalism is being exposed. *Irish Times*. October 10. Available at: https://www.irishtimes.com/opinion/anti-democratic-nature-of-us-capitalism-is-being-exposed-1.894183 (Accessed May 3, 2018).

Choose Wisely. 2018. Choose Wisely Campaign, an initiative of the ABIM Foundation. Available at: http://www.choosingwisely.org/clinician-lists/#keyword=imaging. (Accessed August 12, 2018).

Clarfield, A.M. Gordon, M. Markwell, H. et al. 2003. Ethical issues in end-of-life geriatric care: The approach of three monotheistic religions – Judaism, Catholicism, and Islam. *J Am Geriatr Soc* 51:1149–1154.

Clarke, R.H. 2003. Changing philosophy in ICRP: The evolution of protection ethics and principles. *Int J Low Radiat* 1:39–49.

Coeytaux, K. Bey, E. Christensen, D. et al. 2015. Reported radiation overexposure accidents worldwide, 1980–2013: A systematic review. *PLoS One* 10(3). doi:10.1371/journal.pone.0118709

Cooke, C. 2014. Difference between justice and fairness. Available at: http://www.medicina.uson.mx/wp-content/uploads/2014/02/DIFFERENCE-BTW-JUSTICE-AND-FAIRNESS-Cook.pdf (Accessed June 6, 2018).

Coughlin, S.S. 2008. How many principles for Public Health Ethics? *Open Public Health J* 1:8–16.

Covello, V.T. 2003. Best practices in public health risk and crisis communication. *J Health Comm* 8:5–8.

CPQR (Canadian Partnership for Quality Radiotherapy). 2016. Technical quality control guidelines for Canadian Radiation Treatment Centres. Available at: http://www.cpqr.ca/wp-content/uploads/2017/01/TQC-2016-05-01.pdf (Accessed June 5, 2018).

Curry, A.H. Forastiere, A.A. Flood, W.A. et al. 2014. Overutilization of IMRT/IGRT in treatment of rectal cancer: Cost implications of deviation from evidence based practices. *J Clin Oncol* 32(30_suppl):34–34.

EC. 1997. European Commission. Council Directive 97/43/Euratom of 30 June 1997, on health protection of individuals against the dangers of ionizing radiation in relation to medical exposure, and repealing Directive 84/466/Euratom. *Off J Eur Commun* 180:22–27. Replaced in 2013 by (EC 2013).

EC. 1998a. European Commission. *Radiation Protection. RP 100. Guidance for Protection of Unborn Children and Infants Irradiated Due to Parental Medical Exposures*. Luxembourg: European Commission. DG Environment, Nuclear Safety and Civil Protection. Available at: http://ec.europa.eu/energy/nuclear/radiationprotection/doc/publication/100_en.pdf (Accessed January 21, 2013).

EC. 1998b. European Commission. *Radiation Protection. RP 99. Guidance on Medical Exposures in Medical and Biomedical Research*. Luxembourg: DG Environment, Nuclear Safety and Civil Protection. Available at: http://ec.europa.eu/energy/nuclear/radiation_protection/doc/publication/099_en.pdf (Accessed January 21, 2013).

EC. 2008. European Commission. *RP118. Referral Guidelines for Imaging*. Luxembourg: European Commission, DG TREN, EC. Available at: http://ec.europa.eu/energy/nuclear/radioprotection/publication/doc/118_en.pdf Cited 10 Aug 2009. This publication may no longer be available.

EC. 2009. European Commission. *RP159. European Commission Guidelines on Clinical Audit for Medical Radiological Practices (Diagnostic Radiology, Nuclear Medicine and Radiotherapy)*. Luxembourg: European Commission, DG TREN. Available at: http:/ec.europa.eu/energy/nuclear/radiation_protection/doc/publication/159.pdf (Accessed January 21, 2013).

EC. 2011. European Commission. *Radiation Protection*. RP 167. International Symposium on non-medical imaging exposures. *Proceedings of the Symposium*, Dublin, October 8–9, 2009. Luxembourg: European Commission, DG Energy, 2011. Available at: https://ec.europa.eu/energy/sites/ener/files/documents/167.pdf (Accessed August 1, 2015).

EC. 2012. European Commission. *RP 162. Criteria for Acceptability of Medical Radiological Equipment used in Diagnostic Radiology, Nuclear Medicine and Radiotherapy*. Faulkner, K. Christofides, S. Malone, J. et al. editors. Luxembourg: European Commission. Available at: http:/ec.europa.eu/energy/sites/ener/files/documents/rp162web.pdf (Accessed April 19, 2016).

EC. 2013. European Commission. *Council Directive 2013/59/Euratom: Basic Safety Standards for Protection Against the Dangers Arising from Exposure to Ionising Radiation*. Available at: https://eur-lex.europa.eu/legal-content/EN/TXT/PDF/?uri=CELEX:32013L0059&from=EN (Accessed June 6, 2018). Preceded by (EC 1997).

EC. 2015a. European Commission. *OPERRA (Open Project for Radiation Research Area)*. Available at: http://www.melodi-online.eu/operra.html (Accessed August 1, 2015).

EC. 2015b. European Commission. *RP 180. Medical Radiation Exposure of the European Population, Part 1*. Luxembourg: European Commission, DG Energy. Available at: https://ec.europa.eu/energy/sites/ener/files/documents/RP180.pdf, Survey of DRL's in Part 2. Available at: https://ec.europa.eu/energy/sites/ener/files/documents/RP180%20part2.pdf (Accessed March 28, 2018).

EC. 2018. European Commission. *Radiation Protection Series Publications*. Extensive series of publications, many touching on medical topics. Available at: https://ec.europa.eu/energy/en/radiation-protection-publications (Accessed May 26, 2018).

Edelstein, L. 1943. *The Hippocratic Oath: Text, Translation, and Interpretation*. Baltimore, MA: Johns Hopkins University Press.

Emami, B. Lyman, J. Brown, A. 1991. Tolerance of normal tissue to therapeutic irradiation. *Int J Radiat Oncol Biol Phys* 21:109–122.

EPA. 2009. Environmental Protection Agency. *The Design of Diagnostic Medical Facilities where Ionising Radiation is Used: A Code of Practice* issued by the Radiological Protection Institute of Ireland (now the Office of Radiological Protection). Available at: https://www.epa.ie/pubs/advice/radiation/RPII_Code_Design_Medical_Facilities_09.pdf (Accessed March 28, 2018).

ESR. 2018. European Society of Radiology. See EUROSAFE. Available at: http://www.eurosafeimaging.org/about (Accessed May 26, 2018).

Eurosafe. 2018. EuroSafe imaging together—For patient safety. Available at: http://www.eurosafeimaging.org/about (Accessed May 26, 2018).

Fan, R. 1997. Self-determination vs. family-determination: Two incommensurable principles of autonomy. *Bioethics* 11:309–322.

Fagan, A. 2004. Challenging the bioethical application of the autonomy principle with multicultural societies. *J Appl Philos* 21:15–31.

Faulkner, K. Malone, J.F. Corbett, R.H. et al. 2001. Radiation during pregnancy. In: *Conference: Radiological Protection of Patients in Diagnostic and Interventional Radiology, Nuclear Medicine and Radiotherapy*, pp. 507–511. Vienna, Austria: IAEA. Available at: https://www-pub.iaea.org/books/iaeabooks/6357/Radiological-Protection-of-Patients-in-Diagnostic-and-Interventional-Radiology-Nuclear-Medicine-and-Radiotherapy (Accessed June 6, 2018).

Faulkner, K. Zoetelief, J. Schultz, F.W. eds. 2008. Session on ethics and justification. In Delft Conference Proceedings: Safety and efficacy for new techniques and imaging using new equipment to support European Legislation. *Radiat Prot Dosim* 129:295–310.

Fazel, R. Harlan, M. Krumholz, H.M. et al. 2009. Exposure to low-dose ionizing radiation from medical imaging procedures. *N Engl J Med* 361(9):849–857.

Fox, R. 1990. The evolution of American bioethics: A sociological perspective. In: *Social Science Perspective on Medical Ethics*, ed. Weisz, G., pp. 201–220. Philadelphia, PA: University of Pennsylvania Press.

Funtowicz, S. Ravetz, J. 2003. International society for ecological economics. Post-normal science. *Encyclopedia of Ecological Economics* Available at: http://leopold.asu.edu/sustainability/sites/default/files/Norton,%20Post%20Normal%20Science,%20Funtowicz_1.pdf (Accessed June 6, 2018).

Gardiner, S. 2008. Why we need more than justification in the ethics of nuclear protection: A view from outside. In: *Ethics and Radiological Protection*, eds. Eggermont, G. Feltz, B., pp. 97–111. Louvain-la-Neuve, Belgium: Academia.

Gibbons, M. 2010. *The New Production of Knowledge: The Dynamics of Science and Research in Contemporary Societies.* Los Angeles, CA: Sage Publishing.

GLOBOCAN. 2012. Ferlay, J. Soerjomataram, I. Ervik, M. et al. *Cancer Incidence and Mortality Worldwide*: IARC CancerBase No. 11 v1.0 [Internet]. Lyon, France: International Agency for Research on Cancer; 2013. Available at: http://globocan.iarc.fr/Pages/fact_sheets_population.aspx (Accessed June 6, 2018).

GMC (General Medical Council). 2008. *Lay Membership Statutory Instruments. The General Medical Council (Constitution) Order 2008, No. 2554*. Available at: http://www.opsi.gov.uk/si/si2008/uksi_20082554_en_1 and http://www.gmc-uk.org/about/council/index.asp (Accessed February 7, 2010).

González, A.J. 2011. The Argentine approach to radiation safety: Its ethical basis. *Sci Techn Nucl Inst*. doi:10.1155/2011/910718

Grantzau, T. Thomsen, M.S. Væth, M. et al. 2014. Risk of second primary lung cancer in women after radiotherapy for breast cancer. *Radiother Oncol* 111:366–373. doi:10.1016/j.radonc.2014.05.004

Guenin, M.A. Bigongiari, L. Hernanz-Schulman, M. et al. 2014. ACR practice guideline for imaging pregnant or potentially pregnant adolescents and women with ionizing radiation (Update). Available at: http://www.acr.org/~/media/ACR/Documents/PGTS/guidelines/Pregnant_Patients.pdf (Accessed July 24, 2015).

Günalp, M. Gülünay, B. Polat, O. et al. 2014. Ionising radiation awareness among resident doctors, interns, and radiographers in a university hospital emergency department. *Radiol Med* 119:440–447.

Habermas, J. 1998. *The Postnational Constellation*. Cambridge, MA: MIT Press.

Hager, B. Kraywinkel, K. Keck, B. et al. 2015. Integrated prostate cancer centers might cause an overutilization of radiotherapy for low-risk prostate cancer: A comparison of treatment trends in the United States and Germany from 2004 to 2011. *Radiother Oncol* 115:90–95. doi:10.1016/j.radonc.2015.02.024

Hansson, S.O. 2007. Ethics and radiation protection. *J Radiol Prot* 27:147–156.

Hendee, W. 2013. Risks of medical imaging. *Med Phys* 40:1–2. doi:10.1118/1.4794923

Hendee, W. O'Connor, M.K. 2012. Radiation risks of medical imaging: Separating fact from fantasy. *Radiology* 264(2):312–321. doi:10.1148/radiol.12112678

Hisschemöller, M. Hoppe, R. 1995. Coping with intractable controversies: The case for problem structuring in policy design and analysis. *Knowl Policy* 8(4):40–60. doi:10.1007/BF02832229

Horner, K. 2018. IAEA Webinar: Justification of X-ray examinations in dentistry. Available at: https://www.iaea.org/resources/video/justification-of-x-ray-examinations-in-dentistry (Accessed May 26, 2018).

Horton, P.W. 2011. Dose and risk: The hard facts. In: *Justification of Medical Exposure in Diagnostic Imaging: Proceedings of an International Workshop Brussels*, September 2–4, 2009, pp. 83–90. Vienna, Austria: IAEA. Available at: http://www-pub.iaea.org/MTCD/Publications/PDF/Pub1532_web.pdf (Accessed January 21, 2013).

Huq, M.S. Fraass, B.A. Dunscombe, P.B. et al. 2016. The Report of Task Group 100 of the AAPM: Application of the risk analysis methods to radiation therapy quality management. *Med Phys* 43:4209. doi:10.1118/1.4947547

HPA (Health Protection Agency). 2009a. *Protection of Pregnant Patients during Diagnostic Medical Exposures to Ionizing Radiation*. HPA, RCR, and College of Radiographers RCE-9. Available at: https://www.gov.uk/government/uploads/system/uploads/attachment_data/file/335107/RCE-9_for_web.pdf (Accessed October 31, 2015).

IAEA. 2000. International Atomic Energy Agency. Safety Report Series No.17. *Lessons Learned from Accidental Exposures in Radiotherapy.* Vienna, Austria: IAEA. Available at: https://www-pub.iaea.org/MTCD/Publications/PDF/Pub1084_web.pdf (Accessed June 5, 2018).

IAEA. 2001. International Atomic Energy Agency. Radiological protection of patients in diagnostic and interventional radiology, nuclear medicine and radiotherapy. In: *Proceedings of an International Conference.* Málaga, Spain, March 26–30. Vienna, Austria: IAEA. Available at: https://www-pub.iaea.org/MTCD/Publications/PDF/Pub1113_scr/Pub1113_scr1.pdf (Accessed March 28, 2018).

IAEA. 2009. International Atomic Energy Agency. Report of a consultation on justification of patient exposures in medical imaging. *Rad Prot Dos* 135:137–144.

IAEA. 2011. International Atomic Energy Agency. *Justification of Medical Exposure in Diagnostic Imaging: Proceedings of an International Workshop.* Vienna, Austria: IAEA. Available at: http://www-pub.iaea.org/MTCD/Publications/PDF/Pub1532_web.pdf (Accessed August 1, 2015).

IAEA. 2014. International Atomic Energy Agency. *Radiation Protection and Safety of Radiation Sources: International Basic Safety Standards (BSS). General Safety Requirements.* Vienna, Austria: IAEA. Available at: http://www-pub.iaea.org/MTCD/Publications/PDF/Pub1578_web-57265295.pdf (Accessed February 10, 2016).

IAEA. 2015. International Atomic Energy Agency. *Radiation Protection in Medicine: Setting the Scene for the Next Decade. Proceedings of an International Conference Held in Bonn, Germany.* Vienna, Austria: IAEA. Available at: https://www-pub.iaea.org/MTCD/Publications/PDF/Pub1663_web.pdf (Accessed March 28, 2018).

IAEA. 2018. International Atomic Energy Agency. *RPoP (Radiation Protection of Patients).* Available at: https://www.iaea.org/resources/rpop (Accesses March 28, 2018).

ICRP. 1959. *Recommendations of the International Commission on Radiological Protection.* Publication 1. Oxford, UK: Pergamon Press.

ICRP. 1966. *Recommendations of the International Commission on Radiological Protection.* Publication 9. Oxford, UK: Pergamon Press.

ICRP. 1973. *Implications of Commission Recommendations that Doses be Kept as Low as Readily Achievable.* Publication 22. Oxford, UK: Pergamon Press.

ICRP. 1977. *Recommendations of the International Commission on Radiological Protection.* Publication 26. Annals of the ICRP 1.

ICRP. 1991. *Recommendations of the ICRP.* Annals of the ICRP 20/21, Publication 60. Oxford, UK: Pergamon Press.

ICRP. 2000a. *Pregnancy and Medical Radiation.* Annals of the ICRP 30 (1), Publication 84. Oxford, UK: Pergamon Press.

ICRP. 2000b. *Prevention of Accidents to Patients Undergoing Radiation Therapy.* Annals of the ICRP, 30 (3), Publication 86.

ICRP. 2007a. *The 2007 Recommendations of the International Commission on Radiological Protection.* Annals of the ICRP 37, Publication 103.

ICRP. 2007b. *Radiological Protection in Medicine*. Annals of the ICRP 2007, Publication 105;37(6):1–63.

ICRP. 2018. Cho, K. Cantone, M.C. Kurihara-Saio, C. et al. *Ethical Foundations of the System of Radiological Protection*. Annals of the ICRP 47(1), Publication 138.

IEC. 2018. *International Electrotechnical Commission*. Current list of the IEC Standards for equipment for radiotherapy, nuclear medicine and radiation dosimetry. Available at: http://www.iec.ch/dyn/www/f?p=103:22:13946842664780::::FSP_ORG_ID,FSP_LANG_ID:1362,25 (Accessed May 26, 2018).

Image Gently. 2018. *Image Gently: Pediatric Radiology & Imaging | Radiation Safety*. Available at: https://www.imagegently.org/ (Accessed May 26, 2018).

Image Wisely. 2018. *Image Wisely: Radiation Safety in Adult Medical Imaging*. Available at: https://www.imagewisely.org/ (Accessed May 26, 2018).

Intergovernmental Panel on Climate Change (IPCC). 2014. *IPCC Fifth Assessment Synthesis Report*. Available at: http://ar5-syr.ipcc.ch/ (Accessed July 25, 2018).

IRPA (International Radiation Protection Association). 2014. *Guiding Principles for Establishing a Radiation Protection Culture*. Available at: http://www.irpa.net/docs/IRPA%20Guiding%20Principles%20on%20RP%20Culture%20(2014).pdf (Accessed March 18, 2018).

Jaffray, D. Gospodarowicz, M. 2015. Radiation therapy for cancer. In: *Cancer: Disease Control Priorities*, 3rd ed., Vol. 3, eds. Gelband, H. Jha, P. Sankaranarayanan, R. et al. Chapter 14. Washington, DC: The International Bank for Reconstruction and Development/The World Bank. Available at: https://www.ncbi.nlm.nih.gov/books/NBK343621/ (Accessed May 21, 2018).

James, J.T. 2013. A new, evidence-based estimate of patient harms associated with hospital care. *J Patient Saf* 9:122–128. doi:10.1097/PTS.0b013e3182948a69

Jameton, A. 2010. Environmental health ethics. In: *Environmental Health: From Global to Local*, ed. Frumkin, H., pp. 195–226. San Francisco, CA: John Wiley & Sons.

Jasanoff, S. ed. 2004. *States of Knowledge: The Co-Production of Science and the Social Order*, 1st ed. London, UK: Routledge.

Johnston, D. 2011. *A Brief History of Justice*. Chichester, UK: Wiley-Blackwell.

Jordan, A. Turnpenny, J. 2015. *The Tools of Policy Formulation: Actors, Capacities, Venues and Effects*. Cheltenham, UK: Edward Elgar Publishing.

Justo, L. Villarreal, J. 2003. Autonomy as a universal expectation: A review and a research proposal. *J Asian Int Bioethics* 13:53–57.

Kant, I. 1785. *Grundlegung zur Metaphysik der Sitten* (transl. 1981, *Groundwork of the Metaphysics of Morals*. Cambridge, MA: Hackett Publishing Company.

Kant I. 1795. *Zum Ewigen Frieden. Ein philosophischer Entwurf* (transl. 2003, *Perpetual Peace: A Philosophical Sketch*). Cambridge, MA: Hackett Publishing Company.

Keijzers, G.B. Britton, C.J. 2010. Doctors' knowledge of patient radiation exposure from diagnostic imaging requested in the emergency department. *Med J Aust* 193:450–453.

Kilbrandon, L. 1975. Ethics and the professions. *J Med Ethics* 1:2–4.

Kimura, R. 2014. Japan, bioethics. In: *Bioethics*, 4th ed., Vol. 4, ed. Jennings, B., pp. 1757–1766. Farmington Hills, MI: Macmillan.

Kitcher, P. 2011. *Science in a Democratic Society*. Amherst, NY: Prometheus Books.

Kitcher, P. 2014. *The Ethical Project*. Cambridge, MA: Harvard University Press.

Kolko, J. 2014. *Well-Designed: How to Use Empathy to Create Products People Love*. Cambridge, MA: Harvard Business Review Press.

Kretzmer, D. Klein, E. eds. 2002. *The Concept of Human Dignity in Human Rights Discourse*. The Hague, the Netherlands: Kluwer Law International.

Krille, L. Hammer G.P. Merzenich, H. Zeeb, H. 2010. Systematic review on physician's knowledge about radiation doses and radiation risks of computed tomography. *Eur J Radiol* 76(1):36–41.

Kumar, S. Shelley, M. Harrison, C. et al. 2006. Neo-adjuvant and adjuvant hormone therapy for localised and locally advanced prostate cancer. *Cochrane Database Syst Rev* 18:CD006019. doi:10.1002/14651858.CD006019.pub2

Küng, H. Kuschel, K.J. eds. 1993. *A Global Ethic. The Declaration of the Parliament of the World's Religions*. London, UK: SCM Press.

Lambert, T.W. 1999. Public health risk communication: Ethical considerations. In: *Ethics in a New Age*, Vol. II, ed. Dossetor, J.D., pp. 113–138. Edmonton, Canada: University of Alberta Press.

Lee, C.I. Haims, A.H. Monico, E.P. et al. 2004. Diagnostic CT scans: Assessment of patient, physician, and radiologist awareness of radiation dose and possible risks. *Radiology* 231(2):393–398. See also at: https://www.nwhn.org/getting-burned-radiation-exposure-from-ct-scans/ (Accessed May 31, 2018).

Locke, J. 1689. *Two Treatises of Government*. London, UK: Churchill.

Malone, J. 1998. Justification of elective X-ray exposures in women of childbearing age. In: *Conference: Justification in Radiation Protection*, eds. Faulkner, K. Teunen, D., pp. 7–11. London, UK: British Institute of Radiology. Available at: https://www.researchgate.net/publication/236217112_Justification_of_Elective_X-Ray_Exposures_in_Women_of_Childbearing Age (Accessed June 1, 2018).

Malone, J.F. 2008. New ethical issues for radiation protection in diagnostic radiology. *Radiat Prot Dosimetry* 129:6–12.

Malone, J.F. 2009. Radiation protection in medicine: Ethical framework revisited. *Radiat Prot Dos* 135:71–78. doi:10.1093/rpd/ncp010

Malone, J. 2011a. Justification and tools for change: Scene setting. In: *Justification of Medical Exposure in Diagnostic Imaging: Proceedings of an International Workshop*, pp. 17–24. Vienna, Austria: IAEA. Available at: http://www-pub.iaea.org/MTCD/Publications/PDF/Pub1532_web.pdf (Accessed August 1, 2015).

Malone, J. 2011b. Justification, dose limits and dilemmas. In: *EC RP 167, International Symposium on Non-Medical Imaging Exposures. Proceedings of the Symposium*, pp. 101–112. Luxembourg: DG Energy. http://ec.europa.eu/energy/nuclear/radiation_protection/doc/publication/167.pdf (Accessed August 1, 2015).

Malone, J. 2013. Ethical issues in clinical radiology. In: *Social and Ethical Aspects of Radiation Risk Management.* eds. Oughton, D. Hansson, S.O., pp. 105–130. New York City: Elsevier Science.

Malone, J. 2014. Strategies for improving justification—radiation protection of patients. In: *International Conference on Radiation Protection in Medicine—Setting the Scene for the Next Decade*, pp. 45–50. Vienna, Austria: IAEA. Available at: https://www-pub.iaea.org/MTCD/Publications/PDF/Pub1663_web.pdf (Accessed June 7, 2018).

Malone, J. Guleria, R. Craven, C. et al. 2012. Justification of diagnostic medical exposures, some practical issues: Report of an International Atomic Energy Agency Consultation. *Br J Radiol* 85:523–538. doi:10.1259/bjr/42893576

Malone, J. O'Connor, U. Faulkner, K. eds. 2009. Ethical and justification issues in medical radiation protection. *Radiat Prot Dosimetry (Special Issue)* 135(2).

Malone, J. Perez, M. Godske-Friberg, E. et al. 2016. Justification of CT for individual health assessment of asymptomatic persons: A World Health Organization consultation. *J Am Coll Radiol* 13:1447–1457.

Malone, J. Perez, M. Van Bladel, L. et al. 2015. Clinical imaging guidelines: Risks, benefits, barriers, and solutions. *J Am Coll Radiol* 12(2):158–165. doi:10.1016/j.jacr.2014.07.024

Malone, J.F. Zölzer, F. 2016. Pragmatic ethical basis for radiation protection in diagnostic radiology. *Brit J Radiol* 89(1059). doi:10.1259/bjr.20150713

Matsuoka, E. 2007. The issue of particulars and universals in bioethics: Some ideas from cultural anthropology. *J Philos Ethics Health Care Med* 2:44–65.

Meskens, G. 2016a. Ethics of radiological risk governance: Justice of justification as a central concern. *Ann ICRP* 45(1 Suppl):322–344. doi:10.1177/0146645316639837

Meskens, G. 2016b. Overcoming the framing problem—a critical-ethical perspective on the need to integrate social sciences and humanities and stakeholder contributions in EURATOM radiation protection research. *J Radiol Prot* 36(2):S1–S7. doi:10.1088/0952-4746/36/2/S1

Meskens, G. 2017. Better living (in a complex world): An ethics of care for our modern co-existence. In: *Ethics of Environmental Health*, pp. 115–136. Oxford, UK: Routledge (Studies in Environment and Health, Routledge).

Mill, J.S. 1861. *Utilitarianism*. London, UK: Parker, Son, and Bourn.

Mottet, N. Peneau, M. Mazeron, J.J. et al. 2012. Addition of radiotherapy to long-term androgen deprivation in locally advanced prostate cancer: An open randomised phase 3 trial. *Eur Urol* 62:213–219. doi:10.1016/j.eururo.2012.03.053

MP (Medical Physics). 2013. On the risk to low doses (<100 mSv) of ionizing radiation during medical imaging procedures: IOMP policy statement. *J Med Phys* 38(2):57–58. doi:10.4103/0971-6203.111307

National Archives. 2009. The Shipman report. Available at: http://webarchive.nationalarchives.gov.uk/20090808155110/http://www.the-shipman-inquiry.org.uk/reports.asp (Accessed March 18, 2018).

NCI (National Cancer Institute). 2012. Radiation risks and pediatric computed tomography (CT): A guide for health care providers. Available at: https://www.cancer.gov/about-cancer/causes-prevention/risk/radiation/pediatric-ct-scans (Accessed May 26, 2018).

NCRP. 2009. Publication 160. *Ionizing Radiation Exposure of the Population of the United States*. Bethesda, MD: National Council on Radiation Protection and Measurements. Available at: http://www.ncrppublications.org/Reports/160 (Accessed August 12, 2009).

NCRP. 2011. Publication 170. *Second Primary Cancers and Cardiovascular Disease after Radiation Therapy*. Bethesda, MD: NCRP Publications.

NCRP. 2018. Commentary 27. Implications of recent epidemiologic studies for the linear-nonthreshold model and radiation protection. Available at: www.ncrppublications.org/Commentaries/27 (Accessed May 26, 2018).

NCS. 2013. Nederlandse Commissie voor Stralingsdosimetrie Report 22. Code of practice for the quality assurance and control for intensity modulated radiotherapy. Available at: https://radiationdosimetry.org/?worker=add_footer&text=The+NCS+report+has+been&file=files/documents/0000009/29-ncsreport22imrt-qa.pdf (Accessed June 5, 2018).

Neale, G. Woloshynowych, M. Vincent, C. 2001. Exploring the causes of adverse events in NHS hospital practice. *J R Soc Med* 94(7):322–330.

Nussbaum, M. 2004. Beyond the social contract: Capabilities and global justice. *Oxford Dev Stud* 32:3–16.

Orth, G. ed. 2002. *Die Erde – Lebensfreundlicher Ort für Alle. Göttinger Religionsgespräch 2002 zur Umwelt- und Klimapolitik*. Münster, Germany: LIT Verlag.

Papanicolas, I. Woskie, L.R. Jha, A.K. 2018. Special communication: Health care spending in the United States and other high-income countries. *JAMA* 319(10):1024–1039. doi:10.1001/jama.2018.1150.

Parikh, R.R. Grossbard, M.L. Harrison, L.B. Yahalom, J. 2016. Association of intensity-modulated radiation therapy on overall survival for patients with Hodgkin lymphoma. *Radiother Oncol* 118:52–59. doi:10.1016/j.radonc.2015.10.022

Parliament House of Commons. 2011. Complaints and litigation – Health Committee, 6th report. Available at: https://publications.parliament.uk/pa/cm201012/cmselect/cmhealth/786/78607.htm (Accessed May 21, 2018).

Parsa-Parsi, R.W. 2017. The revised declaration of Geneva: A modern-day physician's pledge. *JAMA* 318(20):1971–1972. doi:10.1001/jama.2017.16230

Pelligrino, E.D. 2008. Some personal reflections on the "appearance" of bioethics today. *Studia Bioetica* 1:52–57.

Persson, L. Shrader-Frechette, K. 2001. An evaluation of the ethical principles of the ICRP's radiation protection standards for workers. *Health Phys* 80:225–234.

PHE/HPA. 2010. Guidance on the safe use of dental cone beam CT (computed tomography) equipment. HPA-CRCE-010. Updated 2015. Available at: https://www.gov.uk/government/publications/dental-cone-beam-computed-tomography-safe-usage (Accessed May 26, 2018).

Picano, E. 2004a. Sustainability of medical imaging. *Br Med J* 328:578–580.

Picano, E. 2004b. Informed consent and communication of risk from radiological and nuclear medicine examinations: How to escape from a communication inferno. *Br Med J* 329(7470):849–851. doi:10.1136/bmj.329.7470.849

Polychroniou, C. 2016. The legacy of the Obama Administration: An interview with Noam Chomsky. Truthout Interview, June 2. Available at: http://www.truth-out.org/news/item/36260-a-mixed-story-ranging-from-criminal-to-moderate-improvement-noam-chomsky-on-obama-s-legacy (Accessed May 2, 2018).

Rathor, M.Y. Rani, M.F.A. Shah, A.S.B.M. et al. 2013. The principle of autonomy as related to personal decision making concerning health and research form an 'Islamic Viewpoint'. *J Islam Med Ass North Am* 43:27–34.

Rawls, J. 1971. *A Theory of Justice.* Cambridge, MA: Harvard University Press.

RCR. 2017. Royal College of Radiologists. *iRefer: Making the Best Use of Clinical Radiology.* London, UK: RCR. Available at: https://rcr-irefer.myshopify.com/?key=34330bb3b22eec7bb482c4ce272efb8fdc71699f74eb4d364ceae40c2984c5ba (Accessed June 8, 2018).

Reilly, D. 2006. A plea for relevance to daily practice. *FOCUS* Fall:18–20.

Renn, O. 2008. *Risk Governance: Coping with Uncertainty in a Complex World.* New York: Earthscan.

Rembielak, A. Woo, T.C. 2005. Intensity-modulated radiation therapy for the treatment of pediatric cancer patients. *Nat Clin Pract Oncol* 2(4):211.

RICOMET. 2018. Risk perception, communication and ethics of exposures to ionising radiation. RICOMET conferences explore the contributions that the humanities and social sciences can make to radiation protection. Available at: http://ricomet2018.sckcen.be/ (Accessed March 18, 2018).

Schaefer, O. 2018. Presenters or patients? A crucial distinction in individual health assessments. *Asian Bioeth Rev* 10(1):67–73.

Schreiner-Karoussou, A. 2008. Review of existing issues and practices with respect to irradiation of patients and staff during pregnancy. *Rad Prot Dos* 129:299–302.

Schreiner-Karoussou, A. 2009. A preliminary study of issues and practices concerning pregnancy and ionising radiation. *Rad Prot Dos* 135:79–82.

Schröder-Bäck, P. Duncan, P. Sherlaw, W. Brall, C. Czabanowska, K. 2014. Teaching seven principles for public health ethics: Towards a curriculum for a short course on ethics in public health programmes. *BMC Med Ethics* 15:73–82.

Semelka, R.C. Armao, D.M. Elias, J. Picano, E. Jr. 2012. The information imperative: Is it time for an informed consent process explaining the risks of medical radiation? *Radiology* 262(1):15–18. doi:10.1148/radiol.11110616

Semghouli, S. Amaoui, B. El Kharras, A. et al. 2017. Physicians knowledge of radiation risk in prescribing CT imaging in Moroccan hospitals. *Curr J Appl Sci Technol* 20(3):1–8. doi:10.9734/BJAST/2017/32491

Sen, A. 2006. *Identity and Violence: The Illusion of Destiny.* New York: W.W. Norton.

Sen, A. 2009. *The Idea of Justice*. Cambridge, MA: Harvard University Press.

SENTINEL. 2008. Conference session with 15 papers on Quality Assurance and Quality Control of radiological imaging devices. *Rad Proc Dos* 129:227–294.

Shah, D.J. Sachs, R.K. Wilson, D.J. 2012. Radiation-induced cancer: A modern view. *Br J Radiol* 85:e1166–e1173. doi:10.1259/bjr/25026140

Shiralkar, S. Rennie, A. Snow, M. et al. 2003. Doctors' knowledge of radiation exposure: Questionnaire study. *Br Med J* 327(7411):371–372. doi:10.1136/bmj.327.7411.371.

Shrader-Frechette, K. Persson, L. 1997. Ethical problems in radiation protection. *Health Phys* 73:373–382.

Sia, S. 2010. *Ethical Contexts and Theoretical Issues: Essays in Ethical Thinking*. Newcastle upon Tyne, UK: Cambridge Scholars Publishing.

Singh, P. Aggarwal, S. Singh Kapoor, A.M. et al. 2015. A prospective study assessing clinicians attitude and knowledge on radiation exposure to patients during radiological investigations. *J Nat Sci Biol Med* 6:398–401. doi:10.4103/0976-9668.160019

SNMMI (Society of Nuclear Medicine and Molecular Imaging). 2018. SNMMI and safe/beneficial medical uses of radiation. Available at: http://www.snmmi.org/ClinicalPractice/content.aspx?ItemNumber=4825 (Accessed May 31, 2018).

Statista. 2015. Number of examinations with computer tomography (CT) in selected countries. Available at: http://www.statista.com/statistics/283085/computer-tomography-examinations-in-selected-countries/ (Accessed September 16, 2015).

Streffer, C. Witt, A. Gethmann, C.F. Heinloth, K. Rumpff, K. 2005. *Ethische Probleme einer langfristigen globalen Energieversorgung*. Berlin, Germany: de Gruyter.

ten Have, H. Gordijn, B. 2013. Global bioethics. In: *Compendium and Atlas of Global Bioethics*, eds. ten Have, H. Gordijn, B., pp. 1–16. Dordrecht, the Netherlands: Springer.

Thomas, E.J. Brennan, T. 2001. Errors and adverse events in medicine: An overview. In: *Clinical Risk Management: Enhancing Patient Safety*, 2nd ed., ed. Vincent, C.A. London, UK: BMJ Publications.

Travis, L.B. Ng, A. Allan, J.M. et al. 2012. Second malignant neoplasms and cardiovascular disease following radiotherapy. *J Natl Cancer Inst* 104(5):357–370.

Tsai, D.F.C. 1999. Ancient Chinese medical ethics and the four principles of biomedical ethics. *J Med Ethics* 25:315–321.

Tsai, D.F.C. 2005. The biomedical principles and Confucius' moral philosophy. *J Med Ethics* 31:159–163.

UN. 1992. United Nations Conference on Environment and Development. Rio declaration on environment and development. Available at: http://www.un-documents.net/rio-dec.htm, Principle 15 (Accessed July 30, 2015).

United Nations General Assembly. 1948. Universal declaration of human rights. Available at: http://www.un.org/en/documents/udhr/index.shtml (Accessed September 16, 2015).

United Nations Conference on the Human Environment. 1972. Declaration on human environment. Available at: http://www.unep.org/Documents.Multilingual/Default.asp?documentid=97&articleid=1503 (Accessed September 16, 2015).

United Nations General Assembly. 1959. Declaration of the rights of the child. Available at: http://www.unicef.org/malaysia/1959-Declaration-of-the-Rights-of-the-Child.pdf (Accessed September 16, 2015).

UNESCO. 1997. Universal declaration on the human genome and human rights. Available at: http://unesdoc.unesco.org/images/0010/001096/109687eb.pdf (Accessed September 16, 2015).

UNESCO. 2005. Universal declaration on bioethics and human rights. Available at: http://unesdoc.unesco.org/images/0014/001461/146180e.pdf (Accessed September 16, 2015).

UNSCEAR (United Nations Scientific Committee on the Effects of Atomic Radiation). 2012. Report of the United Nations Scientific Committee on the effects of atomic radiation; 59th session (General Assembly Official Records. 67th session, May 21–25, 2012). Supplement No. 46. Available at: http://daccess-dds-ny.un.org/doc/UNDOC/GEN/V12/553/85/PDF/V1255385.pdf?OpenElement (Accessed July 17, 2015).

Wagner, L.W. Lester, R.G. Saldana, L.R. 1997. *Exposure of the Pregnant Patient to Diagnostic Radiations: A Guide to Medical Management*. Madison, WI: Medical Physics Publication.

Webley, S. 1996. The interfaith declaration. Constructing a code of ethics for international business. *Bus Ethics: Eur Rev* 5:52–54.

Wennberg, J.E. Brownlee, S. Fisher, E.S. et al. 2008. Improving quality and curbing health care spending: Opportunities for the Congress and the Obama administration. A Dartmouth Atlas White Paper. New Hampshire: Dartmouth Institute for Health Policy and Clinical Practice. http://www.dartmouthatlas.org/downloads/reports/agenda_for_change.pdf (Accessed September 5, 2010).

WHO (World Health Organisation). 2006. *Constitution of the World Health Organization: Basic Documents*, 45th ed., Supplement. Available at: http://www.who.int/governance/eb/who_constitution_en.pdf (Accessed May 6, 2016).

WHO (World Health Organisation). 2014. WHO | Electromagnetic fields and public health: Mobile phones. Available at: http://www.who.int/mediacentre/factsheets/fs193/en/ (Accessed June 8, 2018).

WHO (World Health Organisation). 2015. WHO | Global status report on road safety 2015. WHO. 2015. Available at: http://www.who.int/violence_injury_prevention/road_safety_status/2015/en/ (Accessed June 2018).

WHO (World Health Organisation). 2018a. Tobacco. Available at: http://www.who.int/news-room/fact-sheets/detail/tobacco (Accessed June 2018).

WHO (World Health Organisation). 2018b. WHO global initiative on radiation protection in health care settings. Available at: http://www.who.int/ionizing_radiation/medical_radiation_exposure/global-initiative/en/ (Accessed March 27, 2018).

Wilson, D.S. 2015. *Does Altruism Exist? Culture, Genes, and the Welfare of Others.* New Haven, CT: Yale University Press.

Wilson, G.B. 2008. *Clericalism: The Death of Priesthood.* Collegeville, MN: Liturgical Press.

Wingspread Conference. 1998. The wingspread statement on the precautionary principle. Available at: http://sehn.org/wingspread-conference-on-the-precautionary-principle/ (Accessed August 16, 2018).

World Commission on Environment and Development. 1987. Our common future. Available at: http://www.un-documents.net/our-common-future.pdf (Accessed September 16, 2015).

World Medical Association. 2018. Declaration of Geneva. Available at: https://www.wma.net/policies-post/wma-declaration-of-geneva/ (Accessed November 5, 2017).

World Nuclear Association. 2015. *World Nuclear Power Reactors and Uranium Requirements.* Available at: http://www.world-nuclear.org/info/Facts-and-Figures/World-Nuclear-Power-Reactors-and-Uranium-Requirements/ (Accessed September 16, 2015).

Zölzer, F. 2013. A cross-cultural approach to radiation ethics. In: *Social and Ethical Aspects of Radiation Risk Management*, eds. Oughton, D. Hansson, S.O., pp. 53–70. Oxford, UK: Elsevier Science.

Zölzer, F. 2016. Are the core values of the radiological protection system shared across cultures? *Ann ICRP* 45S:358–372.

Zölzer, F. 2017. A common morality approach to environmental health ethics. In: *Ethics of Environmental Health*, eds. Zölzer, F. Meskens, G., Chapter 4. London, UK, Routledge.

Zölzer, F. Stuck, H. 2018. *Cost-Benefit and Cost-Effectiveness Considerations in the Assessment of Environmental Health Risks: Ethical Aspects.* London, UK: Routledge (in press).

Appendix

AUTHOR CREDITS

This book is the work of the authors. All authors contributed in different ways by developing and refining the concept, identifying the chapter headings, and defining the agenda/content of each chapter. Each chapter was assigned to one of the four as the lead author who, as well as leading the writing of that chapter, was responsible for inviting comments from the other authors as appropriate. As a result, the content evolved dramatically – and sometimes – unpredictably, as we moved along. Whole chapters disappeared and reappeared, sections of one chapter migrated to another. Repetition between and within chapters was identified and minimised, and the level and style was coordinated so that the result is a book with a coherent approach within the limits achievable. This applies to Chapters 1 through 6.

Chapter 7 is somewhat different because it provides a wider reflection on ethics in professional areas involving the public in uncertain sciences. It builds on and extends the understanding achieved in each of the other chapters. It also creates an awareness of approaches and topics that are seldom visited, and examines thoughts that will be valuable for the foreseeable future. The topics involved are not often encountered in discussions on radiation protection and medicine but are, nonetheless, vitally important to a deeper understanding of both. Many readers will find the approach in this chapter both challenging and refreshing.

The lead authors are as follows:

- Preface: Jim Malone and all
- Chapter 1: Jim Malone
- Chapter 2: Friedo Zölzer
- Chapter 3: Jim Malone
- Chapter 4: Jim Malone
- Chapter 5: Christina Skourou
- Chapter 6: Friedo Zölzer
- Chapter 7: Gaston Meskens
- Afterword: Gaston Meskens, Christina Skourou and all

REVIEWERS AND ADVICE

We had the benefit of important and significant external advice on Chapters 4 and 5, dealing with scenarios in medical imaging and radiotherapy, respectively. We extend our gratitude to:

Chapter 4: Dr Denis Remedios, London, UK

Chapter 5: Dr Orla McArdle and Dr Brian O'Neil, Dublin, Ireland

We are extremely grateful to all for their time, patience and insights.

ACKNOWLEDGEMENTS

Jim Malone is grateful to the chairman and trustees of the Robert Boyle Foundation for their ongoing support. The participants in the EC SENTINEL project and the staff/leadership of the International Atomic Energy Agency (IAEA) RPoP section provided fertile environments for fruitful discussions on some of the issues that appear in this book.

Writing this book received no direct funding but drew on experience in various employments and projects in which the authors have been involved.

Index

Note: Page numbers in bold and italics refer to tables and figures, respectively.